FULL OF LIFE

UNESCO Biosphere Reserves – Model Regions for Sustainable Development

Imprint

Published by:
Deutsches MAB-Nationalkomitee beim Bundesministerium für Umwelt, Naturschutz und Reaktorsicherheit (BMU) [German MAB National Committee at the Federal Ministry for the Environment, Nature Conservation and Nuclear Safety]
Chairman: Alfred Walter
Robert-Schuman-Platz 3, D-53175 Bonn
Tel.: (+49 1888) 305-0, Fax: (+49 1888) 305-3225
MAB-Geschäftsstelle im Bundesamt für Naturschutz (BfN) [MAB Secretariat in the Federal Agency for Nature Conservation]
Jürgen Nauber, Birgit Heinze
Konstantinstrasse 110, D-53179 Bonn
Tel.: (+49 228) 8491-0, Fax: (+49 228) 8491-200
http://www.MAB-Germany.info

This publication was produced on behalf of and with funds from the Federal Ministry for the Environment, Nature Conservation and Nuclear Safety.

Project Coordination and Editing:
Birgit Heinze, MAB-Secretariat in the Federal Agency for Nature Conservation, Bonn (Project Management)
Richard Marxen, Thorsten Meyer and Stefan Bröhl, M&P – Partner für Öffentlichkeitsarbeit und Medienentwicklung GmbH, Sankt Augustin

Layout and Graphics:
AD DAS WERBETEAM Werbeagentur und Verlagsgesellschaft mbH, Sankt Augustin

Title Photo: Joachim Jenrich, Gersfeld

ISBN 3-540-20077-0 Springer Berlin Heidelberg New York

Springer is a part of Springer Science+Business Media
springeronline.com

Printed in Germany

Printed on recycled paper (made of 100% waste paper, without colour brighteners)

30/3141/as 5 4 3 2 1 0

As at: December 2003

1st edition
4,000 copies

Proposed bibliography reference:
German MAB National Committee (Ed.) (2005): Full of Life, Bonn.

Forewords

Foreword by the Director-General of UNESCO

Regions and local areas are increasingly important to humankind in times of globalisation. Being able to identify with the area in which they live gives people a sense of belonging and direction, and satisfies the human need for a familiar environment of manageable dimensions. In times of rapid growth and constant change, local involvement enables people to contribute directly and actively to decision-making. This explains in part the increased interest in a success of regional development. Indeed, regionalisation is complementary.

At the same time, global sustainable development has become a key goal for national authorities at the very highest level since the UN Conference on Environment and Development in Rio in 1992. Sustainable development – the balance of ecological, economic and socio-cultural elements, taking into account the needs of future generations in today's decision-making processes – must first be achieved and demonstrated on a more local level.

One of the earliest initiatives to address this issue has been the UNESCO Programme Man and the Biosphere (MAB), with its world network of biosphere reserves and principles of voluntary participation.

One of the most important tasks for the MAB Programme is the development of the biosphere reserve concept. A biosphere reserve is a combination of cultural and natural landscapes that are representative of a country or region, with certain areas designated for nature conservation and others that are managed sustainably. The MAB concept actively incorporates the people living and working in these areas into the further development of the region. Biosphere reserves, therefore, are model regions of sustainable development that are structured in the same way and based on the same principles all over the world. Accordingly, biosphere reserves represent not only different eco-systems but also the broad spectrum of different cultures and economic practices around the world. There are currently 440 UNESCO biosphere reserves in 97 different countries within this worldwide network.

The MAB Programme and its biosphere reserves not only provide suitable research areas and attract highly qualified multidisciplinary scientists, they also offer a committed local population and over 30 years' experience in implementing and testing projects in the area of sustainable development.

I am pleased that Germany – as a highly industrialised country – is committed to developing and testing models for sustainable living and economic practices. This initiative by the German MAB National Committee is warmly welcomed by the MAB community and by the whole of UNESCO. As a diplomat active for many years in promoting economic co-operation for development and in protecting the world's heritage, I also have a great personal interest in the initiative. Publicizing the MAB Programme and the services offered by the UNESCO biosphere reserves to a wider audience, both in German-speaking regions and – with the publication of an English version – internationally, is a further important step. I would like to offer a special word of thanks to the German MAB National Committee and all the scientists involved. Above all, I would like to applaud the understanding, commitment and efforts of the people living in the German biosphere reserves. This book is a valuable contribution to the further development of the UNESCO Man and the Biosphere Programme in Germany and in the world.

Koïchiro Matsuura
Director-General of UNESCO
Paris 2003

Foreword by the Publisher

Dear Readers

How do you want to live in the future – in five, ten or twenty years? What do you wish for your children's lives? Your answer is sure to include safe jobs, a liveable environment, cultural diversity, high environmental quality, attractive landscapes and development opportunities, both personally and for the region in which you live.

There are many blueprints for the future. Often they are too theoretical and involve the people they affect much too little. Since 1971 the UNESCO Programme Man and the Biosphere (MAB) has claimed that it designs and tests models for future development with local people involved. Throughout the world, different paths are followed in 440 model regions, which UNESCO calls "biosphere reserves". This leads to solutions that are both innovative and follow traditions that have proved their worth locally and that can often be transferred to other regions. Very often, these solutions function as an important basis for political decisions because they give equal consideration to ecological, economic and social aspects in an exemplary fashion.

Fourteen areas in Germany belong to the World Network of Biosphere Reserves. The German MAB National Committee – reappointed by the Federal Minister for the Environment, Nature Conservation and Nuclear Safety in the year 2000 – has mainly worked on the conceptual further development of the MAB Programme at national and international level and has also periodically reviewed the German biosphere reserves on behalf of UNESCO. In this book we are going to portray the current state of development in the individual areas, visions and very concrete ideas as well as the potentials of the MAB Programme and our biosphere reserves for shaping the future. We would like to reach a broad readership with this book and we have therefore designed it as "scientific reading" or "readable science". All in all, more than 60 authors have taken part in creating this book. They reveal the large variety of players involved in implementing the MAB Programme. In their contributions they give their own opinions, views and experience. The articles in this book are just as different and diverse as the biosphere reserves themselves.

We thank all of the authors for their commitment, which contributed to the success of the project. This publication was planned and realised in less than a year. This was only possible due to the enthusiasm, the elan and the great dedication of all involved. The MAB Programme is "full of life"! The work on this book has impressively proved this to us and whetted our appetite for the future.

Naturally, the compilation of this book was associated with considerable editorial and coordination work due to the vast difference of the articles and the large number of people involved. We would therefore like to thank Thorsten Meyer and Stefan Bröhl from the agency "M&P – Partner für Öffentlichkeitsarbeit und Medienentwicklung GmbH" for their committed editorial work. Our special thanks go to Birgit Heinze from the Secretariat of the German MAB National Committee, who took on the organisation of the entire project with a great deal of enthusiasm and tremendous dedication.

German MAB National Committee
Bonn 2003

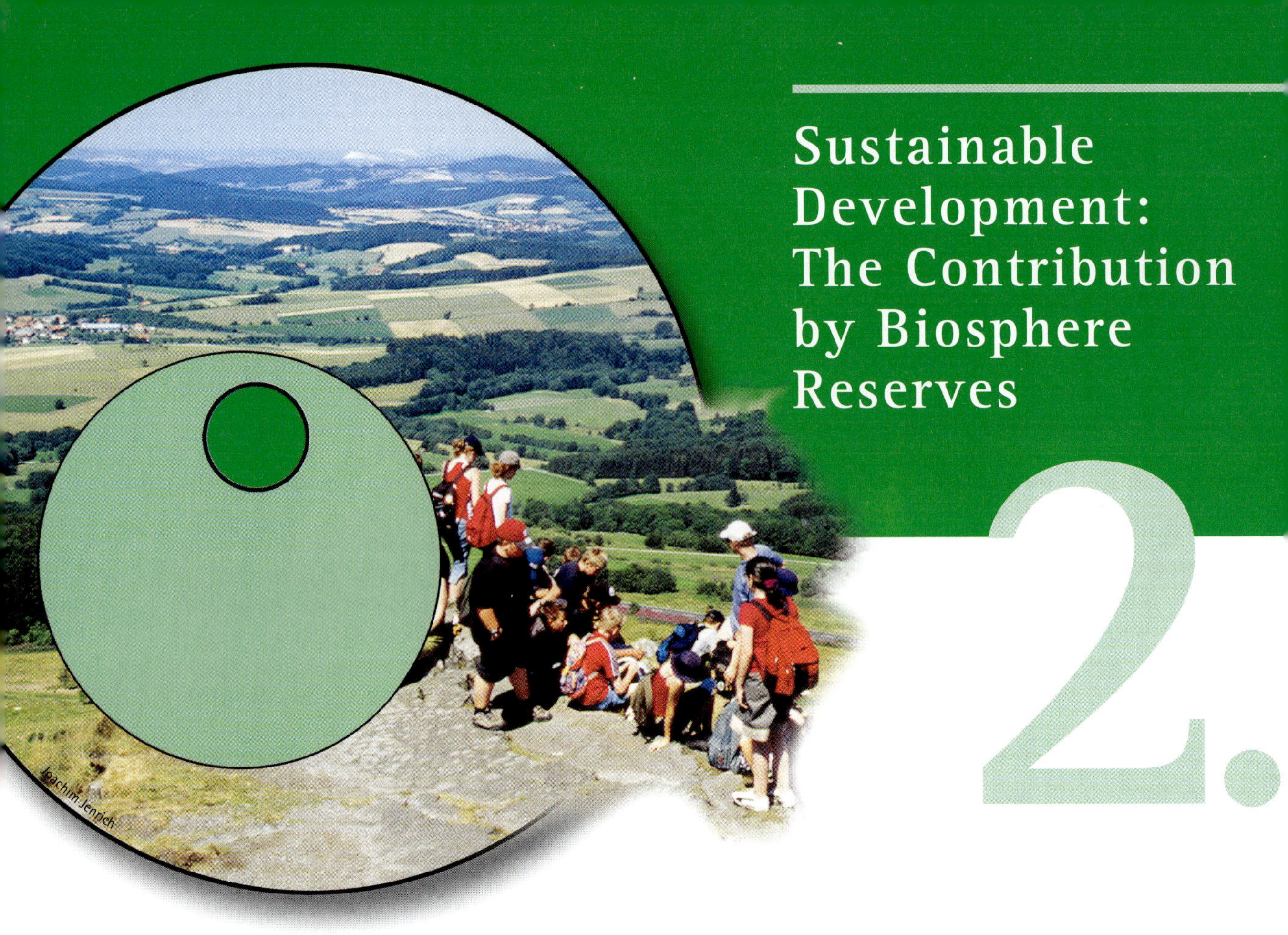

Sustainable Development: The Contribution by Biosphere Reserves

2.

2.1 *MAB – a Programme over the Course of Time*

Alfred Walter, Folkert Precht and Rolf-Dieter Preyer

The UNESCO Programme Man and the Biosphere (MAB) was established in 1970. It started out as a purely scientific programme and over time has grown into a world network of model regions for sustainable development (cf Chapter 2.2).
In the early days, the programme objective was to acquire the fundamental scientific principles required at an international level for the protection of natural resources and for an environmentally compatible use of the biosphere. The MAB Programme was therefore the first international environmental programme focusing on the relationship between humans and the environment.
Nearly all UNESCO member states started national implementation immediately after the launch of the Programme. By setting up MAB national committees, the Federal Republic of Germany and the German Democratic Republic in 1972 and 1974 respectively, fulfilled an essential formal requirement for participation in the MAB Programme.
As an applied research programme, it quickly became clear that it needed special instruments to turn the results of the research into political action. The World Network of Biosphere Reserves was therefore established in 1976 (cf Chapter 2.2).
Following the UN Conference on Environment and Development (UNCED) in Rio de Janeiro in 1992 the participating states in the MAB Programme increased their focus on sustainable development. By virtue of their concept, the biosphere reserves should be predestined to contribute reasonably to executing the decisions made at the UNCED Conference, such as the implementation of Agenda 21 and the Convention on Biological Diversity (CBD).
The MAB Programme was granted its current conceptual foundation in the Seville Strategy, adopted by the UNESCO General Conference in 1995 (28C/Resolution 2.4).
The International Guidelines for the World Network of Biosphere Reserves, agreed at the same time, established a new institutional framework for the World Network, binding in form and content. As a result, every biosphere reserve has to comply with a series of minimum conditions before it is included in the World Network. Nature and landscapes must be protected, economic and human development promoted and environmental education, training, research and monitoring supported. The involvement of the local population is imperative to this.
The International Guidelines for the World Network lay down compulsory criteria for the recognition and periodic review of biosphere reserves. Every ten years the condition of the biosphere reserves should be reviewed on the basis of these criteria.
Following the first review of biosphere reserves in Germany in 2001, the German MAB National Committee established that "sustainable life systems and sustainable economic

development" in UNESCO biosphere reserves had been neglected at a national and international level up to this point – despite sustainable development being the focus of the MAB Programme.

The national committee regards it as particularly important to develop biosphere reserves as model regions for sustainable regional development. A highly industrialised country such as Germany has a special responsibility within the World Network to develop and test sustainable ways of life and economic systems and quality economies.

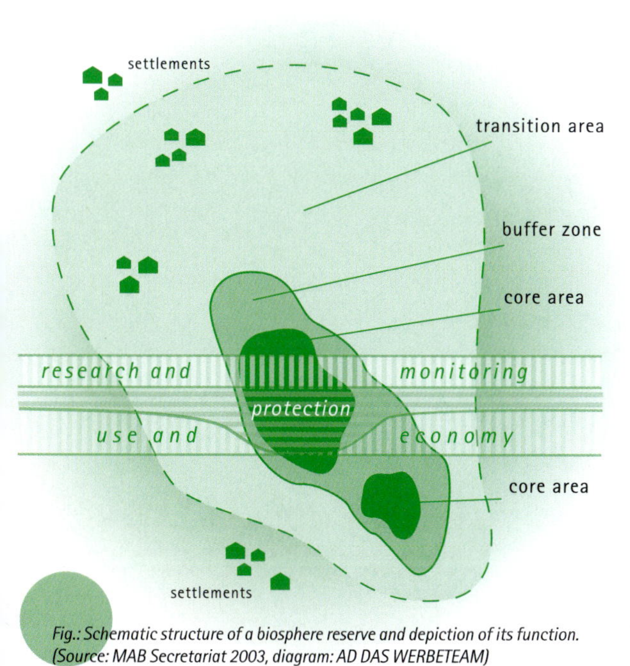

Fig.: Schematic structure of a biosphere reserve and depiction of its function. (Source: MAB Secretariat 2003, diagram: AD DAS WERBETEAM)

Literature

UNESCO (Ed.) (1996): Biosphere Reserves. The Seville Strategy and the Statutory Framework of the World Network, Paris.

2.2 *World Network of Biosphere Reserves*

Jürgen Nauber

Biosphere reserves are the main instrument of the UNESCO Programme Man and the Biosphere (MAB). As of August 2003, 97 countries from over 140 participating states have designated a total of 440 biosphere reserves.

The Statutory Framework of the World Network of Biosphere Reserves and the Seville Strategy (1995) established a worldwide network from the many individual areas (UNESCO 1996). The Statutory Framework was approved by the UNESCO General Conference in 1995 and forms the legal basis for the biosphere reserves, without being binding under international law. However, they embody far more the principle of a voluntary approach to cooperation. By cooperating with one another, the states are committing themselves to accepting the criteria and guidelines of the MAB Programme. Biosphere reserves do not only use conventional methods to protect valuable ecosystems in their core areas, such as national parks. Much more, they also make it possible and call for a sustainable economy in the transition area of the biosphere reserve. Through the Worldwide Network of Biosphere Reserves UNESCO is making an important instrument available to the international community for the national implementation of the results of the UN Conference on Environment and Development (UNCED) in Rio de Janeiro in 1992 and the Convention on Biological Diversity (CBD).

The Statutory Framework lays down a specific procedure for the recognition of biosphere reserves. In addition, every ten years the condition of each biosphere reserve is examined by an independent committee of experts using the criteria of the Statutory Framework and individual objectives set for each area. As a result, recommendations and suggestions for improvement are made which support the states in their efforts to develop biosphere reserves.

The World Network of Biosphere Reserves is coordinated by the UNESCO MAB Secretariat in Paris. The threads of the individual national MAB structures come together there.

The MAB Secretariat organises meetings, looks after the flow of information within the network (cf www.unesco.org/mab), coordinates studies, provides assistance with technical issues and advises on all matters relating to biosphere reserves.

The collaborators see themselves as "brokers" for the biosphere reserves and arrange financing and establish contacts. In addition, the MAB Secretariat represents the World Network when dealing with other institutions and organisations. It represents the World Network at events and conferences and when working with the secretariats of conventions and other international programmes.

2. SUSTAINABLE DEVELOPMENT: THE CONTRIBUTION BY BIOSPHERE RESERVES

In recent years, more and more biosphere reserves that extend across national borders have been recognised and registered by UNESCO. This shows that biosphere reserves also facilitate political relations. The protection and the sustainable use of connected landscapes, separated "only" by political boundaries, has been made possible through the setting up of transboundary biosphere reserves. Areas that are stable from an ecological and economic point of view have been created and relations with neighbouring countries have improved through sustainable regional development. In this way, biosphere reserves can also contribute to preventing crises and solving conflicts (cf Chapter 4.14).

It is not only the number of applications for recognition as biosphere reserves that has increased considerably over the last five years. There has also been a marked improvement in the quality of the applications in terms of the biosphere reserves' contribution to sustainable regional development. This is a result of the adoption of the Statutory Framework

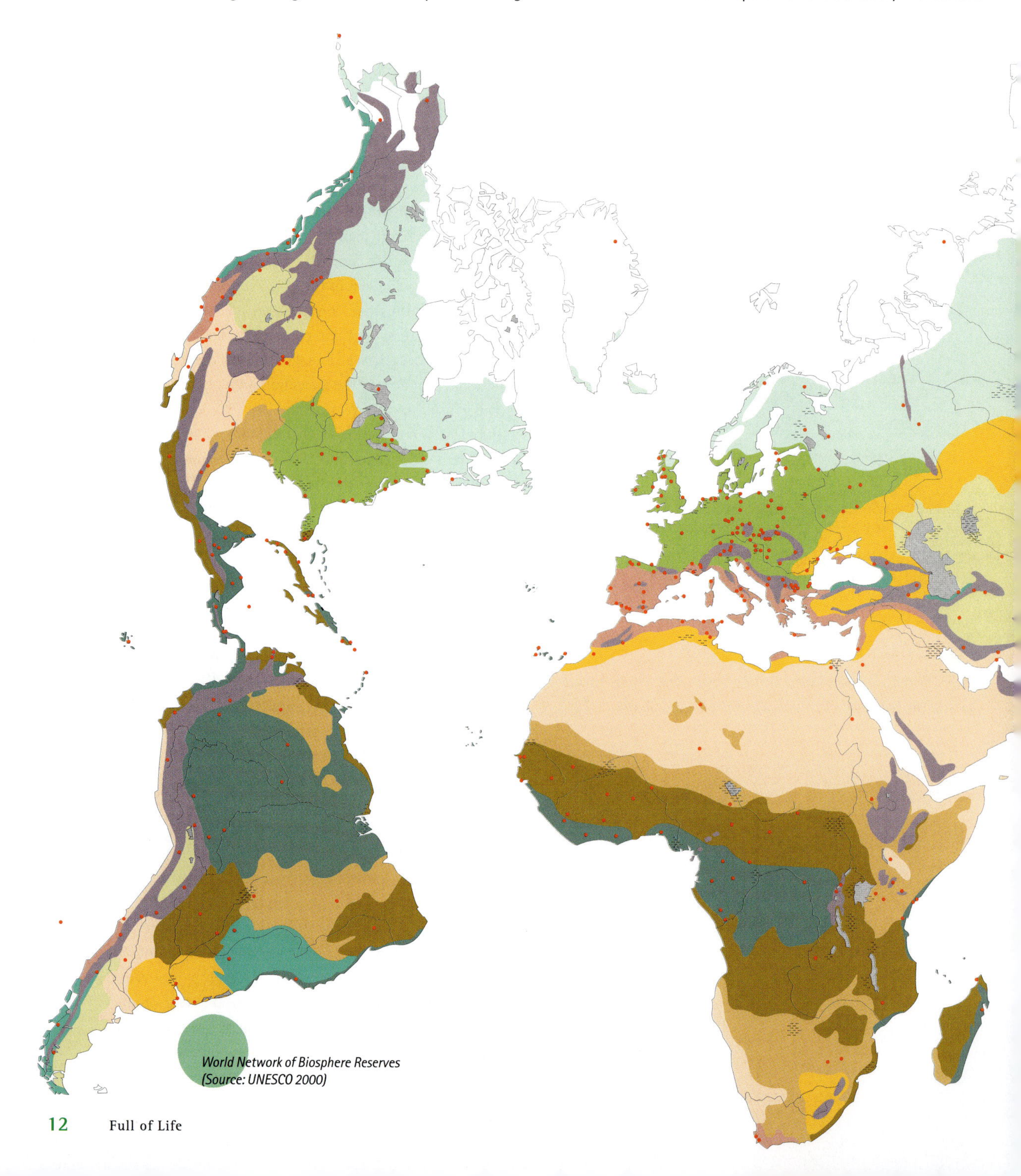

World Network of Biosphere Reserves (Source: UNESCO 2000)

and the Seville Strategy in 1995, which act as kind of guideline to the architect's plan for the successful implementation of the biosphere reserve concept. Also the surface area of the regions applying for recognition has become noticeably larger, as extensive transition areas, chosen on the basis of their economic suitability, are required to fulfil the economic objectives of the MAB Programme.
Despite all the success to date, a lot of work still remains to be carried out to develop the World Network of Biosphere Reserves. Many areas were recognised when nature conservation was the main focus of the MAB Programme. It is now necessary to expand on this so that the Seville Strategy can also be employed. The ecological work has not yet been completed, either. Many ecosystems are not yet sufficiently represented in the World Network, such as mountains, coastlines or deserts. There is also a real need to catch up on work required in many regions of Africa, Asia and South America. The World Network of the Biosphere Reserves will make an important contribution here to the implementation of the recommendations agreed at the UN Conference on Sustainable Development in Johannesburg in 2002.

The World Heritage Convention

In addition to the World Network of Biosphere Reserves, UNESCO has established another network for World Heritage Sites. In the framework of the Convention Concerning the Protection of the World Cultural and Natural Heritage – UNESCO World Heritage Convention (1972), which is an international agreement, natural and cultural landscapes have been identified in addition to cultural sites. Whereas biosphere reserves should be representative of the world's ecosystems, the universal outstanding importance of each of the World Heritage Sites comes to the fore. This is why the World Heritage Convention is much more concerned with preservation, whereas in the biosphere reserves the main focus is worldwide representative nature and development.
Nevertheless, these concepts complement one another. The core area of a biosphere reserve can also be protected as a World Heritage Area at an international level. There are many examples of this worldwide. Examples include the Aggtalek and Slovensky Kras Biosphere Reserves on the Hungarian-Slovakian border, where the chalk caves designated as Natural Heritage Sites are located, or the Palawan Biosphere Reserve in the Philippines, where two national parks have been designated World Heritage Sites and form the core area of the Biosphere Reserve (www.unesco.org/mab/BR-WH.htm).

Literature

UNESCO (1996): Biosphere Reserves: The Seville Strategy and the Statutory Framework of the World Network, Paris.

2.3 Biosphere Reserves: Model Regions for the Future

Harald Plachter, Lenelis Kruse-Graumann and Werner Schulz

MAB: The Programme for the 21st Century

When UNESCO announced a scientific programme called Man and the Biosphere (MAB) in 1971, the response was rather muted. What did it mean? And, furthermore: it was only one of very many international, regional and national research programmes that dealt with the relationship between humans and nature.

Looking back from today's perspective, this programme was the first to consistently place a basic idea at its heart that now – over 30 years later – has become a supreme global guiding principle in politics. At that time, the term "sustainability" did not yet exist as a political programme and, nevertheless, the title "Man and the Biosphere" was precisely what we now understand it to be today. Yet, this programme, just like so many others, would probably have been pushed to the back of a drawer if there hadn't been a second idea; to set up a worldwide network of representative areas where innovative, sensitive forms of nature utilisation were to be developed by research and practice: biosphere reserves. In retrospect, the name may appear unfortunate. "Reserves" are too reminiscent of protected areas that exclude people, oppress local and indigenous cultures and, therefore, not at all of a future-oriented strategy. But, nevertheless, the programme and the term have not only survived; today they are more topical than ever. The heart of the MAB Programme in the early 1970s was not much more than a vague vision in the minds of a few scientists. In politics today it occupies a similar standing to terms like "peace" or "economic stability".

Global Guideline of "Sustainability"

With the "Technical Revolution" of the first half of the 19th century and the findings of modern science that developed over the same period, for the first time in history humans had the means to free themselves from a close, not infrequently vital dependency on nature. The new technologies seemed to be so convincing that no doubts could be raised about their advantages or their long-term viability. Early critics of this technology-credulity, such as the German poet and nature conservationist Hermann Löns, remained lonely "voices in the desert" (cf PLACHTER, H. 1991).

Remarkably, that it is precisely the technology that has probably saved most human lives to date that was the one that for the first time gave rise to fundamental doubts about the limitations of scientific and social development. New types of artificial pesticides, such as DDT, helped millions of people to feed themselves adequately, to successfully fight against crucial threats like malaria, and thus to survive. However, modern ecology, which was developing at the same time, documented shocking effects on nature. Rachel Carson's book "The Silent Spring" (1962) was the first element to shatter an apparently fixed image of the world. An avalanche of reports about more negative effects of modern technology followed, culminating in "Red Data Books of Extinct and Endangered Species", the founding of environment ministries and the first serious political and economic consequences. Our societies have still not got over this cultural shock of the 1960s: undoubtedly, the needs of a rapidly growing world population could be satisfied only with the help of modern technologies and new social structures. Its risks for nature and – through nature – for human health turned out to be much greater than had been thought. Appropriate compromises that go beyond pure bans were hardly in sight and if they were, they appeared to be not realisable politically.

It was not until the second half of the 1980s that this state of affairs was tackled seriously in the political sphere. Among other things, building on a little regarded definition by the World Conservation Union IUCN (then: International Union for the Conservation of Nature and Natural Resources; cf Box 1), an international commission under the leadership of the former Norwegian Prime Minister Brundtland put a new political term at the heart of its considerations (GOODLAND, R. et al. 1992). It took up the principle of "sustainability" as a system of management that "satisfies present needs without compromising the ability of future generations to meet their own needs". Finally, at the UN Conference on Environment and Development (UNCED) in Rio de Janeiro in 1992 the United Nations (UN) declared sustainability a general guiding principle for the 21st century. Since poverty is one of the major reasons for predatory exploitation of nature, it made the global fight against poverty into a central solution strategy. In the UN concept, economic growth and more wealth for all become the locomotive of future viability. "However, integration of environment and development concerns and greater attention to them will lead to the fulfilment of basic needs, improved living standards for all, better protected and managed ecosystems and a safer, more prosperous future..."

Since the Rio World Summit in 1992, the guiding principle of sustainable development has gained a foothold in political institutions and programmes at all levels. For example, the international community has made commitments in joint

Box 1: Basic definition of the "Protection of Biotic Resources" (which means in substance "nature conservation"), by the World Conservation Union IUCN in 1980 (IUCN 1980), slightly abbreviated

(...)

- *to maintain essential ecological processes and life-support systems (such as soil regeneration and protection, the recycling of nutrients, and the cleansing of waters), on which human survival and development depends;*
- *to maintain genetic diversity (...) on which depend the functioning of many of the above processes and life-support systems, the breeding programmes necessary for the protection and improvement of cultivated plants, domesticated animals and microorganisms (...).*
- *to ensure the sustainable utilisation of species and ecosystems (...), which support millions of rural communities as well as major industries.*

agreements such as the Montreal and Kyoto Protocols to protect the ozone layer and the global climate and has advanced the fight against poverty with the Doha Declaration, which is to grant the least developed countries access to worldwide markets.

Competitive Europe

The European Union (EU), too, made sustainable development a central component of its common policy in the 1997 Amsterdam Treaty. At the 2001 Gothenburg Summit, it presented a strategy entitled "A Sustainable Europe for a Better World" that expanded the strategic goals for economic and social policy that had been laid down in Lisbon one year earlier with an ecological dimension. In its strategy, the European Commission cites the protection of the climate and resources as well as the preservation of health and mobility as key points. At the same time, it wants to make "Europe the most competitive and dynamic knowledge-based economy in the world". Under the motto "Global Partnership", there is a separate focus on the external dimension of sustainability – combating worldwide poverty.

Perspectives for Germany

The implementation of the European objective at national level defines the Federal Government's 2002 sustainability strategy under the title "Perspectives for Germany". In this, the Federal Government defines sustainability as an interdisciplinary task that is to be a fundamental principle in its policy in all fields in future. On the whole, the strategy formulates guiding principles of sustainable action for the key areas of energy, transport, health protection and food, family and old age, education and innovation. There is a separate focus on combating poverty, fostering development and worldwide environmental and resource conservation. The recommendation to understand sustainability as a locomotive for innovation and to face up to the challenges of globalisation and structural change with a sustainable way of doing business is addressed at companies (www.bundesregierung.de).

Local Agenda 21

Numerous German local authorities together with several thousand cities and communities throughout the world are on the way towards a local agenda. The trigger for this movement was the final Rio document of 1992, the Agenda 21. This global programme of action for sustainable development was signed with binding effect by most countries on earth – including Germany. The document portrays demands for sustainable development at national and international level. Furthermore, local authorities all over the world are called upon to develop their own programmes of action in the form of "Local Agendas".

By now, agenda processes referring to individual towns and cities have been set in motion in practically all German cities (www.agendaservice.de). In the Scandinavian countries and the United Kingdom, programmes of action of this kind have not just been drawn up for towns and cities, but also for the majority of rural local authorities.

What is Sustainability?

The term "sustainability" is much older than its current popularity would lead us to believe. In fact, the history of sustainability goes back to Saxony in the baroque age. In Freiberg in around 1700, Chief Inspector of Mines Carl von Carlowitz developed a counter model to the severe degradation of forests practised until then. To conserve the wood resources in the long run, he recommended that only so much wood should be felled as could grow back through reforestation. However, a definition of sustainability of this kind, only relating to type and quantities of resources, can no longer meet modern, ecologically-based perceptions of careful use.

Sustainability Triangle "Ecology, Social Affairs & Economy"

In 1992, industrialised and developing countries agreed in Rio on the confirmation of the future goal of global, sustainable development. Since the "Rio plus ten" follow-up conference in Johannesburg (2002) at the latest, this goal has been defined so that it goes beyond the mere maintenance of the ability of the ecological system to function. Much rather, the objective includes the idea – also assuming social, ethical and economic dimensions – of a life with human dignity based on individual self-development, both for the current and future generations. What is essential about this definition is that it understands sustainable development as an interdisciplinary task that basically affects all areas of society equally and that it sets a clearly future-oriented emphasis with responsibility for future generations (BUNDESREGIERUNG 1999, ENQUETE-KOMMISSION 1998, HABER, W. 1998 b).

In this general definition, the term sustainability has experienced very broad social and political approval. However, it is not operable in this form. Consequently, the time after Rio has been characterised by intensive efforts to define the term more precisely and take it into account in decision-making processes. Some things have been achieved, but much has been left open to this day, not only in detail, but also in fundamental issues.

Fig. 2: UNESCO World Heritage cultural landscape "The rice terraces of Ifugao/ Philippines": Sensitive use of nature...

... for future generations (Batad, Philippines)

Box 2: Perspectives for Sustainability

The sustainability approach aims at bringing together economic performance, social responsibility and environmental protection to facilitate fair development opportunities for all countries and to preserve the natural foundations of life for future generations. Currently, throughout the world there are around 70 attempts to bring this guiding principle ("regulative idea") closer to operationalisation. Examples include:

- *If the ecologists have their way, the ecosystems should not be overtaxed by a use of its resources.*
- *Most economists view sustainable development as an economic form that has to ensure that the same welfare will be available for future generations as for those of today.*
- *Physicists call for the conservation of biological systems that are stable within themselves, and chemists would like all anthropogenically influenced substance cycles to be closed where possible (i.e. "recycling").*

Particularly drastic examples of non-sustainable economies are

- *deforestation in the Mediterranean area by the Romans and the destruction of tropical forests today,*
- *overfishing of the oceans by ever more perfect catching techniques and*
- *steppisation of large parts of the former Lake Aral in Russia as a consequence of the diversion of large quantities of water to irrigate agriculture.*

Examples of sustainable economic development are harder to find, especially if not all forms of economic activity that owe their permanence only to the low levels of technical intervention in the past are to be called sustainable. In principle, the following types of economic activity can be considered sustainable:

- *cultivation of centuries old rice terraces in China and South-East Asia,*
- *various forms of agricultural forest use (agro-foresting) in Africa and Latin America and*
- *cultivation of Alpine pastures from the 17th century to the end of the Second World War.*

Open Questions and Attempts to Solve Them

The past discussion on sustainability is overloaded with a number of ideological interpretations that do not stand up to rational examination. Behind this are fundamental values that automatically lead to communication problems if the individual point of view is not indicated sufficiently clearly. Some general related problems refer to:

1. Economics, social sciences and ecology (nature conservation, see below) interpret the term of sustainability in very different ways. To this day there is no really shared basic understanding of what sustainable action could be.

2. Interpretations of the sustainability principle are not infrequently supported by the (not openly expressed) ideal of a "life in harmony with nature". In this connection, frequent references are made to historical cultures or indigenous peoples who would come relatively close to this ideal. Many historical economic systems, however, were by no means ecologically sustainable, but have sometimes even led to the extinction of the culture in question due to predatory exploitation and overuse. In the case of indigenous peoples in obviously semi-natural regions, the main question to be adressed is the extent to which their traditional ways of living and working correspond to modern ideas of social sustainability (cf Chapter 3.1.1).

3. Broad circles, not just in the general public and among politicians, start from the erroneous assumption that there are solutions where economic, social and ecological interests can be given equal maximal consideration at the same place. However, the development of sustainability strategies always means that a compromise or a balance for different interests will be found in a process. On the one hand, possible solutions can be integrating concepts at the same place, but on the other hand the spatial separation of priorities ("priority areas"; integrating and segregating strategies, see below).

4. The specific application of the sustainability principles therefore not least depends on the area levels that are selected. There are no generally valid "patent solutions" that can be applied one-to-one to various local situations. Sustainability concepts for Europe or Germany must be designed differently than those at regional or local authority level and solutions that were developed at one place cannot be transferred to others until they have been adapted.

5. The lack of a precise definition of sustainability is often justified with the lack of scientific data. This is certainly not wrong. However, data, no matter how precise, will never "automatically" lead to useful solutions. Just as important are standardising and thus explaining normative steps, based on value principles, in the form of agreements between various interest groups. Thus, for example, sustainability strategies should also consider the interests of future generations. But to how many generations should this apply and who, in the case of doubt, has priority, the living or the future generations? Scientific data do not provide an answer to this. Solutions must be found between different points of view and interests, and the way or methodology how to achieve these must be understood by everybody. Any form of use, however we imagine it, changes nature and, also, human's social environment. Sustainable development of humankind and "untouched nature" do not go together (irrespective of this, society can consciously decide not to use certain areas, maybe for reasons of nature conservation or because there are no economic perspectives of use). But how much nature is still sustainable and what are the indicators for the relevant nature quantities and qualities? Here, too, research results do not provide an "automatic" decision. Sustainable development must be the result of a comparison of societal values and consensus. This valuing dimension of sustainability still is not treated accordingly in many discussions.

6. Sustainability thus mainly means rethinking values and developing new forms of decision-making for everyday problems. The latter is also necessary because the conventional decision-making processes are optimised to guidelines that cannot be harmonised with the idea of sustainability (e.g. preferring short-term technical progress over the long-term safeguarding of development, making decisions on the basis of scientific facts without considering questions of values, placing individual interests above those of the community and future generations). Agenda 21 initiatives are certainly a pioneering element for this. Yet they alone are not enough. Their efforts will have only few effects as long as they are trapped by the conventional thinking of policy and economics on the use of nature. Sustainability does not thus arise solely from scientific data, but mainly in the hearts and minds of those people who decide about their own futures and those of their children.

The Pillars of Sustainability

Nature Conservation

In the last few years a growing gap has opened up between the public perception and the scientific concept of nature conservation. The main reasons for this are:

1. Increasingly, conventional species and ecosystem conservation is perceived as the sole field of work (discussions on ecosystem mapping, nature conservation areas and the Habitats Directive of the European Union). But nature conservation comprises all natural commodities, including the so-called abiotic ones, such as water and soils. It pursues a nationwide, spatially and thematically differentiated concept of aims and has always included a future-oriented development strategy that considers human usage interests through landscape planning (cf IUCN definitions, Box 1 and Article 1 of the Federal German Nature Conservation Act). The sustainable development of nature and landscapes has long been a central issue for nature conservation (PLACHTER, H. 1991).

2. In recent years, the practical everyday work of nature conservation has increasingly returned to static and preserving protection strategies that tend to look backwards and to biotic-oriented commodities. This is certainly a direct consequence of the continuing losses of and interventions in nature. But this cannot mean that the other fields of activity are neglected in the long term due to a lack of capacities and/or low public acceptance. Above all, there is a lack of new ideas and practicable approaches for these fields of activity (SRU 1996, 2000).

3. Ecology and nature conservation are largely placed on an equal footing in terms of content. However, whereas ecology is an empirical science, nature conservation is a valuing, results-oriented action discipline (ERZ, W. 1986). Ecology and nature conservation have a relationship with each other similar to that between biology and human medicine or between physics or chemistry and the engineering sciences. Consequently, nature conservation needs a broad extra spectrum of methods that ecology does not deliver, e.g. in the fields of value identification, value comparison, decision-making and planning (PLACHTER, H. et al. 2002).

Fig. 2: Historical landscape, Central Swabian Alb 1936: fields in the valley bottom, overgrazed oligotrophic limestone grassland (Source: Schenkel Archive, LfU Baden-Württemberg)

Nature conservation can in no way mean only the conservation of intact nature or nature that has been influenced by humans as little as possible. Agricultural ecosystems also function without any problems in the scientific sense. Much rather, the difference is that they are artificially kept stable by means of constant human influence, especially in the form of energy and substance inputs, and often have greater negative impacts (especially for humans themselves) than natural ones. Nature conservation has several "basic motives", including the protection and development of biodiversity, the stability of natural systems, unique natural creations, the conservation of wild species and natural ecosystems, and the development of systems of use adapted to nature (PLACHTER, H. 1999). For the discussion about sustainability, it is decisive that the character (and thus, ultimately, the "value") of these basic motives are positively related to each other only in very specific cases and in a few places on earth. Some tropical forests and large coral reefs are natural, biologically diverse, stable and unique, all at the same time. In most other cases, the characters of the individual basic motives do not depend on each other. Many natural ecosystems are extremely poor in species and/or are not very stable. In many places on earth – for example in Central Europe – humans have greatly increased biological diversity over time in comparison to the natural state in ways that, by modern standards, are far from being "sustainable" (cf Fig. 1). Early land use forms in Europe were

Fig. 1: A landscape east of Zadar, Croatia that has been overused for 2,000 years. The options for future generations were spoiled.

by no means "extensive" in terms of use of labour and exhaustion of natural resources, as is often claimed today. Nevertheless, many of the ecosystems that arose in that way, such as oligotrophic grasslands, grazed woodland or heaths, are considered to be of prime importance for conservation today due to their high biodiversity and their landscape aesthetics (cf Fig. 2).
This means that there can be no single, uniform "sustainability indicator" for the field of ecology. Which basic motive is to have priority over the others in which place has to be decided in individual cases and ultimately – with all of the help from scientific data – by setting normative standards. Above all, "nature conservation" means maintaining the diversity of nature in all of its aspects. But these differ from place to place. The enormous wealth of nature on this earth is the result of the differences between locations. Politics and the administration, however, aim at general, simple guidelines that can be applied with legal certainty everywhere. It is this in particular that entails the great danger of levelling out differences in locations – and thus lowering diversity – instead of fostering them. "Ecological sustainability" can only be developed in relation to areas and subject matter. A worldwide network of "model regions", in which conservation and development strategies adapted to locations are developed, is the key logistical foundation for this.
The contribution of nature conservation to a sustainability strategy cannot exhaust itself in a sporadic, conserving method of protecting species and semi-natural ecosystems nor in a call for the reintroduction of pseudo-extensive historical forms of land use (note: the majority of so-called contractual nature conservation strategies pursues precisely this goal). Nationwide concepts, pioneering ideas and – above all – a placing of value on natural commodities that is transparent to the public are required. In this sense, environmental protection primarily aimed at human health is ultimately only a partial component of a more comprehensive strategy for the protection of an intact nature (= environment).

Economy

The central task of human economic activity lies in creating economic value by means of entrepreneurial activities. Economic activity, however, is not only for the short-term maximisation of profits, but also for the satisfaction of human needs and, thus, the provision of livelihoods for all individuals (cf box 3). In the long term, the economic component of sustainable development can therefore be described as an economic form that has to ensure that the same welfare will be available for future generations as for those of today. The strategic goal of sustainable development should therefore be to develop products and services for the future markets of a society with a sustainable economy.

Box 3: Economic Approaches for Sustainability

- *Encourage innovations for the development of ecological products and markets!*
- *Cooperate or form networks in the product line or to change the market!*
- *Use the opportunities presented by regional structures by buying materials and products from the region!*
- *Use potentials for cost savings by means of ecological and social measures in the company (e.g. reducing sickness costs)!*
- *Invest in projects that are economically, ecologically and socially meaningful!*
- *Conduct fair competition on the market!*
- *Pay salaries and wages in line with collective bargaining or typical for the sector!*
- *Encourage ecological and social projects, for example by means of donations or sponsoring!*

(Bundesumweltministerium/Umweltbundesamt 2001)

Social Aspects

What is decisive for the standardising process of sustainable development is that none of the three dimensions of ecology, economy or social affairs may be individually optimised, but that a solution should only ever be sought and found involving and considering the other two components.
Whereas there are still relatively rounded provisions for the ecological dimension, as a sustainable, protective and wise use of natural resources, and the economic dimension, as the means of satisfying needs for current and future generations by means of economic development, this is not the case for the social dimension. The core of the social dimension is the safeguarding of equity and equality of opportunity within the generation existing today (e.g. balance between North and South, but also West and East now) and between the present and future generations. This equalisation, also called the intergenerational agreement, concerns equity within a generation in the first case and equity between different generations in the second case. If these aspects are considered, we also talk about "socially compatible" ecological and economic development. However, the social dimension goes far beyond this definition of terms.
There are differences even in the names for the social dimension: it is often called "socio-cultural" in order to emphasise the culturally specific differences and characteristics (e.g. in comparison between the North and the South). In other cases, the cultural dimension is considered to be the "fourth leg" of a chair (cf Chapter 3.1.1), which must not wobble at all. But

the cultural dimension is also seen as a dimension encompassing the three main dimensions because it is cultural schemes, values and practices that structure and link the three dimensions of sustainability and weigh them up against each other. Others add "participation" as the fourth leg and use it to refer to a procedure related to the three contents.
The term "sustainable development" is often criticised as ambiguous and "fuzzy" and a precise definition is demanded. However, it is precisely the relatively broad interpretation scope that offers a way in for many areas of policy and many scientific disciplines. Nevertheless, the concept of sustainability forces the overcoming of sectoral and departmental boundaries and a merger – or at least a debate with – such different scientific disciplines as the natural, economic and social sciences. Without this integrative concept of sustainable development, cooperation of this kind would not come about so easily.
Furthermore, it is essential for the shaping of sustainable development that sustainable development only makes sense as a **global** process that, however, can be realised only **locally** in the region (natural and cultural space). The term "glocalisation" is starting to establish itself for processes of this kind (CHARNIAWSKA, B. 2003). In spite of advancing globalisation, in spite of necessary global framework conditions, it is becoming ever clearer that the local level of action plays a major role, both as the origin for global development and as the location of the impact of global developments on the local population and on natural resources.
The role of people as **shapers and sponsors** of sustainable development will be moved to the fore. Not only biosphere reserve managers, but also and above all the various individuals and interest groups involved (players, stakeholders) in a biosphere reserve must help to make decisions, support and, ultimately, implement the various forms and characteristics of protection and use. The prerequisite for this is that all of these players should be interested and actively involved. This, in turn, presupposes profound knowledge of the predominant individual and socio-cultural values, the subject areas with potential for conflict, the motivation structures, the responsibilities and the conditions for further action that can be effective as obstacles or as potential in the process of social change towards sustainable development.
This means that the social dimension of sustainable development includes all individual, social and culturally specific conditions that are relevant for the human-nature and/or human-environment relationship. This relationship is largely built upon and co-determined by the importance of nature for every individual person. This subjective importance is based on traditional knowledge in society (e.g. indigenous knowledge), individual belief systems (nature cannot be destroyed at all) and collective ideas (neem trees are worshiped in many parts of India and are therefore protected) and it is influenced and changed by continuous social communication, whether direct from person to person or via the media. That is why it makes sense to design biosphere reserves as "social-ecological units".

Vision and Reality

Currently (September 2003) 440 biosphere reserves have been recognised by UNESCO in all parts of the world. What contribution have they made to the basic ideas of the MAB Programme and to the political guiding principle of sustainability? There is still no systematic analysis of what has been achieved in biosphere reserves. The "periodical reports", compiled every ten years for every biosphere reserve according to the MAB Programme and have also been compiled for the German biosphere reserves since 2001, provide information about the level of development. However, it will probably be a few years yet until a worldwide image can be derived from these reports. To date, 97 countries have become involved in the Programme. The biosphere reserves cover an area of approximately 45.1 million square kilometres (425 biosphere reserves, as of June 2003). Almost all of the biosphere reserves have their own staff of state employees.
In 1971, the ideas of the MAB Programme were so innovative that at first there were only vague conceptions of how to realise them. In the early days in particular, therefore, biosphere reserves were often established in outstanding natural areas without any significant human population or land use. Not infrequently, there were existing national parks or even Category I wilderness areas under the World Conservation Union (IUCN), such as the Amboseli National Park in Africa or the Yellowstone National Park in the USA. The recognition of the German national parks of the Bavarian Forest and Berchtesgaden as biosphere reserves also dates back to this time. In the former Soviet Union "biosphere zapovedniki" are a separate statutory category of protected area. The core area of the established areas there is relatively big and very well protected, meaning that the total area often come very close to IUCN category II (national parks).
Nevertheless, the existing system of biosphere reserves is much more than another category of large-scale protected areas. The principles of the Seville Strategy of 1995 once again made this very clear and adapted the MAB Programme to the current discussion about sustainable development (GERMAN MAB NATIONAL COMMITTEE 1996).
All biosphere reserves have spatial zoning, usually comprising a core area, a buffer zone and a transition area. It is especially the strictly protected, unused core area that has repeatedly given cause for misunderstandings on the concept and goals of biosphere reserves. The idea of developing sustainable ways of nature utilisation by people in biosphere reserves is obvious and convincing. But is it also credible if there are simultaneous demands to totally remove a certain proportion of the land from any human use? This could give rise to the suspicion that nature conservationists strive to use

Fig. 3: Rhön Biosphere Reserve: museum landscape or landscape of the future? An additional income for local communities of about € 130 million from tourism speaks a clear language.

biosphere reserves as not much more than another instrument for increasing the number of conventionally protected areas in another way.

Precisely because biosphere reserves are test areas, in other words small worlds where the nature utilisation of the future is tested on a model scale, strictly protected core areas are an essential part of the concept. The scientific discussion on the general nature conservation goals of the last few decades makes a distinction between a segregative and an integrative strategy. In the segregative strategy, nature conservation areas and areas used by humans are separated from each other geographically. In the integrative strategy, existing forms of use should be improved so that nature and use can co-exist at the same location or very close to each other. The literature on this subject is vast, the spatial models are often overlaid by the author's individual understanding of nature and, not least, also by the region in which the ecological analysis used as a basis came about. Works from regions with large nature areas subject to current risks tend towards a segregative strategy, those from old cultural landscapes with constant use more towards an integrative approach.

Today, there can no longer be any doubt that adequate nature conservation in densely populated areas can only be achieved by both strategies together. Since the beginning, the concept of biosphere reserves has pursued a "partial integration strategy" of this kind (PLACHTER, H., REICH, M. 1994). A large number of species and ecosystems can exist only under largely natural conditions and many ecological processes, too, are possible only in unused areas. If we consider larger areas of landscape, as biosphere reserves usually are, a network of mainly unused areas is the skeleton of a functioning natural balance. This idea is at the heart of the European "Natura 2000" Programme, for instance.

But this alone cannot be enough. In many parts of the world, almost unlimited use on the remaining area has led to the ecological disaster that we are facing today. This means that concepts for the integration of technical nature conservation goals in the existing and future forms of nature utilisation as just as necessary as a core area concept. In this connection, a distinction must be made between areas in which nature conservation goals have priority over use interests because of the sensitivity of the natural commodities without the use interests being ruled out in general (the buffer zones of the biosphere reserves; Fig. 3) and others in which use is to the fore (transition areas). However, in generally used areas, too, a minimum degree of nature must be conserved so that they preserve usage options for future generations along the lines of the idea of sustainability.

Thanks to the zoning concept of biosphere reserves, the priorities for protection functions and/or use options are already defined at the basic level. Nevertheless, protection and use are not static terms that have been set in stone, but constantly need to be reconfirmed, respected, adapted, "learnt" by locals, visitors and subsequent generations (cf Chapters 3.1.1 and 3.1.2).

If we use the system of differentiated nature use (HABER, W. 1971, 1998 a; WBGU 2001) "protection ahead of use", "protection through use" and "protection in spite of use" as the basic terms, biosphere reserves are especially suitable for applying these different strategies. It is precisely this close connection between nature conservation (protection ahead of use in the core area), landscape conservation, nature management activities in the buffer zone (protection through use) and sustainable economics in the transition area with low-resource waste, regional marketing and sustainable tourism (protection in spite of use) that will have to be developed even more and advocated as a model (Realisation of the concept of "best practice").

Consequently, the question of "how much nature" is necessary for this has to be raised. Guideline values, such as those found in the Federal Nature Conservation Act (10 per cent of the area), can be indicated in general and for very large areas. However, if such guideline values were transferred to smaller areas without adaptation, this would level out landscape diversity and thus lead to a loss of biodiversity. The buzzword of "Think globally, act locally" applies to hardly any field of applied science and technology more than to nature conservation. Only a broad spectrum of local concepts adapted to the locational stand conditions, history and the development potentials can ultimately ensure the realisation of the global objectives of nature conservation. Exactly for this reason, a worldwide network of model regions, as the biosphere reserves are dedicated to be, is a basic requirement for any viable sustainability strategy.

As a result of this, biosphere reserves undoubtedly occupy a special position: To develop new concepts and technologies, far more expenditure than normal is necessary. This includes the bundling of research and trials in biosphere reserves as well as the creation of special incentives for accepting new things. However, this must not go so far that the resultant areas differ from the surrounding area in key fields of development. Because then the models developed there would no longer be transferable (FISCHER, W. 2000).

The Contribution of German Biosphere Reserves

Fig. 4: Sustainability starts in the minds of our children

The state of development in the 14 German biosphere reserves with regard to their contributions to a sustainability strategy varies. However, on the whole, they are on a very good course on an international scale. Especially in view of the comparatively poor support from the state and of science, the work of the German biosphere reserves must be evaluated as extremely efficient. Nevertheless, there is still need for further action:

- The concept of the MAB Programme is highly topical and convincing in view of the current discussion about sustainability. However, this should be communicated internally and externally much more comprehensively than in the past. Brochures, lectures and information centres are important, but are not enough for this. A comprehensive way of learning for sustainability and the development of new forms of learning (cf Chapter 3.1.2) must become a central task for all biosphere reserves (cf Seville Strategy).
- At state and political level the opportunities presented by the instrument of biosphere reserves are not perceived adequately. The often sparse staffing of individual biosphere reserves is only one sign of this. What is more decisive is that biosphere reserves have hardly played a role to date in any area of state technology and development programmes. If certain technological or economic programmes take place in biosphere reserves, this is usually chance and by no means deliberate bundling. Moreover, it is barely understandable that for many years plenty of biosphere reserves have had problems in identifying a minimum number of areas as core areas although more or less enough state land is available.
- The situation is similar for research that is increasingly financed by so-called external funding via individual project applications due to empty state coffers and for research policy reasons. In spite of all the advantages, this system undoubtedly increases the scientist's dependency on the ideas and guidelines of the external funding body. This implies greater responsibility of funding bodies to give priority to research projects that are important to society. Although sustainability is such an important focus, biosphere reserves do not play a notable role in the research plans of relevant donors, either at European or at German level. In particular, there is a lack of incentives that could motivate a scientist or a group of scientists to locate the study areas of a project in a biosphere reserve.
- Unmistakably, many biosphere reserve administrations are – at least not in the latest development – focused on "classical" nature conservation strategies, including maintenance measures and contractual nature conservation. However, this is automatically the case because, on the one hand, in the start-up phase in particular, sufficient natural potential has to be safeguarded and recreated; many natural commodities cannot be regenerated, or can be regenerated only over very long periods. On the other hand, the administrations not only lack the responsibilities for incentive systems of other areas of society (e.g. agricultural subsidies, tax incentives), but not infrequently they also do not have the personnel technical expertise. The start-up phase of the German biosphere reserve network must now be followed by a phase that brings the structural and economic development of the area (and thus also the transition areas) more to the fore.
- Only a little of what has been newly developed in biosphere reserves has entered into routine processes outside these areas. In this connection, the as yet inadequate documentation of the successes plays a role here, as well as the fact that in the age of globalisation political decision-makers are increasingly ignoring local development progress, even systematically preventing it with some of their activities. Approaches that have recognised this, such as the Council of Europe's Landscape Convention, should also be given greater consideration than in the past in biosphere reserves.
- In many cases the local population does not sufficiently identify with "its" biosphere reserve. A not insignificant aspect of this is bound to be the fact that although many positive developments (e.g. the recreational value and, thus, tourism) have benefited from the biosphere reserve status, it is not possible for the individual citizen to make a connection of this kind.

Literature

BUNDESREGIERUNG (1999): Stichwort "Nachhaltigkeit", Bonn.

CARSON, R. L. (1962): Silent Spring, New York.

CHARNIAWSKA, B. (2003): A Tale of Three Cities, Oxford.

ENQUETE-KOMMISSION "SCHUTZ DES MENSCHEN UND DER UMWELT – ZIELE UND RAHMENBEDINGUNGEN EINER NACHHALTIG ZUKUNFTSVERTRÄGLICHEN ENTWICKLUNG" DES 13. DEUTSCHEN BUNDESTAGS (Ed.) (1998): Konzept Nachhaltigkeit. Vom Leitbild zur Umsetzung. Abschlussbericht, Bonn.

ERZ, W. (1986): Ökologie oder Naturschutz? Überlegungen zur terminologischen Trennung und Zusammenführung. In: Ber.

Akad. Naturschutz Landschaftspfl. 10, pp. 11-17.
FISCHER, W. (2000): Sind Biosphärenreservate Modellregionen für zukunftsfähige Entwicklung? Jülich: www.itas.fzk.de/deu/tadn/tadn002/fisc00b.htm
GERMAN ADVISORY COUNCIL ON GLOBAL CHANGE (WBGU) (2001): World in Transition: Conservation and Sustainable Use of the Biosphere, Berlin.
GERMAN MAB NATIONAL COMMITTEE (Ed.) (1996): Criteria for Designation and Evaluation of UNESCO Biosphere Reserves in Germany, Bonn.
GOODLAND, R., DALY, H., EL SERAFY, S. & B. VON DROSTE (Eds.) (1992): Nach dem Brundtland-Bericht: Umweltverträgliche wirtschaftliche Entwicklung, Bonn.
HABER, W. (1971): Landschaftspflege durch differenzierte Bodennutzung. In: Bayerisches Landwirtschaftliches Jahrbuch, 48 (Sonderheft 1), pp. 19-35.
HABER , W. (1998 a): Das Konzept der differenzierten Landnutzung - Grundlage für Naturschutz und nachhaltige Entwicklung. In: BUNDESUMWELTMINISTERIUM (Ed.): Ziele des Naturschutzes und einer nachhaltigen Naturnutzung in Deutschland - Tagungsband zum Fachgespräch, Bonn, pp. 57-64.
HABER, W. (1998 b): Nachhaltigkeit als Leitbild der Umwelt- und Raumentwicklung in Europa. In: Ber. 51. Dt. Geographentag, Bd. 2, Bonn, pp. 11-31.
INTERNATIONAL UNION FOR THE CONSERVATION OF NATURE AND NATURAL RESOURCES (IUCN) (1980): World Conservation Strategy: Living Resource Conservation for Sustainable Development, Gland (Switzerland).
PLACHTER, H. (1991): Naturschutz, Stuttgart.
PLACHTER, H. (1999): The Contributions of Cultural Landscapes to Nature Conservation. In: BUNDESDENKMALAMT WIEN (Ed.): Monument, Site, Cultural Landscape, Exemplified by the Wachau, Wien, pp. 93-115.
PLACHTER, H., REICH, M. (1994): Großflächige Schutz- und Vorrangräume: Eine neue Strategie des Naturschutzes in Kulturlandschaften. Verhandl. Projekt Angewandte Ökologie (PAÖ) 8, Stuttgart, pp. 17-43.
PLACHTER, H., BERNOTAT, D., MÜSSNER, R. & U. RIECKEN (Eds.) (2002): Entwicklung und Festlegung von Methodenstandards im Naturschutz. Schr. R. Landschaftspflege u. Naturschutz 70, Bonn.
PRESSE- UND INFORMATIONSAMT DER BUNDESREGIERUNG (Ed.) (2002): Perspektiven für Deutschland. Unsere Strategie für eine nachhaltige Entwicklung, Berlin.
SRU (RAT VON SACHVERSTÄNDIGEN FÜR UMWELTFRAGEN) (1996): Konzepte einer dauerhaft umweltgerechten Nutzung ländlicher Räume. Sondergutachten. (Deutscher Bundestag, Drucksache 13/4109).
SRU (RAT VON SACHVERSTÄNDIGEN FÜR UMWELTFRAGEN) (2000): Umweltgutachten 2000. Kurzfassung. http://www.umweltrat.de.
UNESCO (Ed.) (1996): Biosphere Reserves. The Seville Strategy and the Statutory Framework of the World Network, Paris.

2.4 The Network of Biosphere Reserves in Germany

Dieter Mayerl

The development into the current network of biosphere reserves in Germany must be viewed in conjunction with the development of the UNESCO Man and the Biosphere Programme (MAB).
Although UNESCO had recognised the sustainable management of representative landscapes and the people managing them as integral components from the outset (ERDMANN, K.-H. 1997; ERDMANN, K.-H., NAUBER, J. 1995), this cannot be said for the development in Germany: as well as the scientific aspect, the goal of achieving international recognition via the UNESCO biosphere reserves was high on the agenda in Germany. The development of biosphere reserves in Germany therefore is reflecting a piece of German-German history.

The German-German History of the Biosphere Reserves

The Federal Republic of Germany (FRG) and the former German Democratic Republic (GDR) founded National Committees in 1972 and 1974 respectively, thus establishing the conditions for participation in the MAB Programme. In 1979 UNESCO recognised the Steckby-Lödderitz Forest (today part of the Elbe River Landscape Biosphere Reserve in the *Land* Saxony-Anhalt) and the Vessertal (today Vessertal-Thuringian Forest Biosphere Reserve in the *Land* Thuringia) as the first biosphere reserves in the GDR. The first recognition in the Federal Republic of Germany followed in 1981 for the Bavarian Forest. The two German states set research-based emphases when fleshing out the framework of the MAB Programme prescribed by UNESCO (DEUTSCHES MAB-NATIONALKOMITEE 1995).
The biosphere reserves in Germany received particular attention thanks to the decision of the GDR Council of Ministers of 22 March 1990, i.e. after the fall of the Berlin Wall, to set up a National Park Programme. This programme dealt with five national parks and three nature parks as well as four new biosphere reserves (Rhön, Schorfheide-Chorin, Spree Forest and South-East Rügen Biosphere Reserve) as well as the extension of the two biosphere reserves that had already been recognised, Middle Elbe and Vessertal-Thuringian Forest (AGBR 1995).

On 12 September 1990 – shortly before German Unification – the landscapes in the GDR National Park Programme were legally safeguarded. The regulations entered into force on 1 October 1990. Shortly afterwards, UNESCO recognised these areas as biosphere reserves and also recognised the extension of the existing biosphere reserves.

In the west of Germany this was followed by the establishment of additional areas, which were then recognised by UNESCO, for example the Wadden Sea and the Palatinate Forest. In 1996, UNESCO recognised the Upper Lausitz Heath and Pond Landscape in the *Land* Saxony as the 13th biosphere reserve. This area comprises the biggest and ecologically richest pond landscape in Germany – in other words managed areas. They also include approximately 2,000 hectares of former lignite mining land that are to be regenerated.

Finally, the extension of the Middle Elbe BR in Saxony-Anhalt into the Elbe River Landscape BR comprising five *Länder* (Schleswig-Holstein, Mecklenburg-Vorpommern, Lower Saxony, Brandenburg, Saxony-Anhalt) with an area of 342,848 hectares along a 400 kilometre stretch of river in 1997 and the recognition of the Schaalsee BR as the 14th biosphere reserve in the year 2000 mark the increasingly clearer development of the German network of biosphere reserves into model regions of sustainable economic development.

But the development of the network is not yet complete. This concerns both the areas themselves and the number of areas (cf Chapter 3.5).

German Definition of Biosphere Reserves

"Biosphere reserves are large, representative sections of natural and cultural landscapes. They are sub-divided according to the influence of human activity into a core area, a buffer zone and a transition area, which can also contain a regeneration zone. Most of the area of the biosphere reserve should be legally protected.

In biosphere reserves – together with the people living and working there – model concepts on conservation, maintenance and development are drawn up and implemented. At the same time, biosphere reserves help to research relations between man and the environment, ecological environmental monitoring and environmental education. They are recognised by UNESCO within the context of the Man and the Biosphere Programme."

(AGBR 1995: 5)

Overview of Biosphere Reserves in Germany

The 14 German biosphere reserves recognised by UNESCO with a total area of 15,798 square kilometres cover around 4.43 per cent of the area of Germany. Fig. 1 and Tab. 1 give an overview of the location, size and status of the area with the relevant proportions of the area in the zones.

The biosphere reserves in Germany have developed in different ways due to the individual history of their formation, the administrative and, sometimes, legal enshrining in the *Länder* and the relevant financial and staff resources.

Unlike in the eastern German *Länder*, existing protected areas – or parts of them – with the status of national parks and nature parks were recognised as biosphere reserves in the western *Länder*. The approach caused some problems because the areas – often without adequate transition areas – could not be further developed in line with the UNESCO Seville Strategy (UNESCO 1996) and the International Guidelines of the World Network of Biosphere Reserves (UNESCO 1996). A solution still has to be found in these biosphere reserves.

Thanks to the network of German biosphere reserves, the development is largely harmonious and sustainable, taking account of the particular situation in each case. The Permanent Working Group of Biosphere Reserves in Germany (*Ständige Arbeitsgruppe der Biosphärenreservate in Deutschland* AGBR) set up in 1990 (see below) and the Guidelines for Conservation, Maintenance and Development issued by it (AGBR 1995) have made a major contribution to this.

The biosphere reserves in Germany in their entirety broadly represent the major landscapes or landscape types in Germany. The Wadden Sea BR encompass the mudflats, islands and marshes of the German North Sea coast, the South-East Rügen BR the landscape of the Mecklenburg-Vorpommern coast formed in the Ice Age and shaped in the post-Ice Age.

The Schaalsee BR comprises a cultural landscape marked by the Ice Age with deep lakes and bogs rich in lime, the Upper Lausitz Heath and Pond Landscape BR one of the biggest pond areas in Germany. The Schorfheide-Chorin BR represents a complete section of the northern German young moraine landscape. The Elbe River Landscape BR and Spreewald BR comprise plains and glacial valleys in the northern German old moraine landscape, the Rhön, Vessertal-Thuringian Forest and Bavarian Forest Biosphere Reserves various landscape types of the Central European medium range mountains. The German part of the Palatinate Forest-North Vosges BR represents the south-west German stratified land and the Berchtesgaden BR the northern chalk Alps.

Other major landscapes or natural spaces, such as the foothills of the Alps, the north-west German geest, the Rhenish slate mountains or an urban-industrialised area are not yet represented by the biosphere reserves in Germany.

Fig. 1: Biosphere Reserves in Germany (BfN 2002: 126)

Biosphere Reserve	Federal *Land*	UNESCO-Recog-nition	Core Area [ha]	[%]	Buffer Zone [ha]	[%]	Transition Area [ha]	[%]	Total Area [ha]
South-East Rügen	*Mecklenburg-Vorpommern*	*07.03.1991*	*349*	*1.5*	*3,204*	*16.0*	*19,947*	*82.5*	*23,500*
Schleswig-Holstein Wadden Sea*	*Schleswig-Holstein*	*16.11.1990*	*85,500*	*30.0*	*6,400*	*2.2*	*193,100*	*67.8*	*285,000*
Hamburg Wadden Sea	*Hamburg*	*10.11.1992*	*10,530*	*89.7*	*1,170*	*10.3*	-		*11,700*
Lower Saxon Wadden Sea	*Lower Saxony*	*10.11.1992*	*130,000*	*54.2*	*108,000*	*45.0*	*2,000*	*0.8*	*240,000*
Schaalsee	*Mecklenburg-Vorpommern*	*21.01.2000*	*1,709*	*5.5*	*7,905*	*25.8*	*21,286*	*68.9*	*30,900*
Schorfheide-Chorin	*Brandenburg*	*16.11.1990*	*3,648*	*2.8*	*24,103*	*18.7*	*101,410*	*78.5*	*129,161*
Elbe River Landscape (Middle Elbe)	*Brandenburg, Mecklenburg-Vorpommern, Lower Saxony, Schleswig-Holstein, Saxony Anhalt (Saxony Anhalt)*	*15.12.1997 (24.11.1979)*	*7,220*	*2.1*	*61,726*	*18.0*	*273,902*	*79.9*	*342,848*
Spree Forest	*Brandenburg*	*07.03.1991*	*974*	*2.1*	*9,334*	*19.6*	*37,201*	*78.3*	*47,509*
Upper Lausitz Heath and Pond Landscape	*Saxony*	*15.04.1996*	*1,124*	*3.7*	*12,015*	*39.9*	*16,963*	*56.4*	*30,102*
Vessertal–Thuringian Forest	*Thuringia*	*24.11.1979*	*437*	*2.6*	*2,024*	*11.8*	*14,637*	*85.6*	*17,098*
Rhön	*Bavaria, Hesse, Thuringia*	*07.03.1991*	*4,199*	*2.3*	*67,483*	*36.5*	*113,257*	*61.2*	*184,939*
Palatinate Forest-North Vosges (only German part)	*Rhineland-Palatinate*	*10.11.1992*	*3,739*	*2.1*	*49,261*	*27.7*	*124,000*	*70.2*	*177,000*
Bavarian Forest	*Bavaria*	*15.12.1981*	*10,224*	*76.7*	*3,105*	*23.3*	-		*13,329*
Berchtesgaden	*Bavaria*	*16.11.1990*	*13,896*	*29.7*	*6,948*	*14.9*	*25,898*	*55.4*	*46,742*
TOTAL			**273,549**		**362,678**		**943,601**		**1,579,828**

Tab. 1: Biosphere Reserves in Germany
*(Source: MAB Secretariat in the Federal Agency for Nature Conservation (BfN; based on information from the biosphere reserves as of: 30.06.2003; *01.02.2000)*

The spectrum of ecosystems, in particular fauna and flora, represented is just as diverse as the major landscapes comprised by the network of biosphere reserves (cf Table 2). It contains both semi-natural and varying degrees of human-made ecosystems.

All of the biosphere reserves are used – albeit in different ways. Even such apparently pristine landscapes as the Wadden Sea, the Elbe river meadows or the Berchtesgaden Alps are clearly influenced by people. With the exception of the Elbe River Landscape BR the biosphere reserves are in rural, economically peripheral areas. To date, there are no urban-industrial areas.

Due to their comparatively low environmental pollution, their natural resources and the attractive countryside, biosphere reserves are popular holiday destinations. Tourism is one of the most important sources of income and employment for the local population.

There are some major conflicts of use and damage to the natural balance in the biosphere reserves. These conflicts mainly result from the sealing of surfaces and fragmentation of areas caused by residential and industrial areas as well as by traffic and infrastructure, from pollutant inputs, mass tourism or land use not compatible with the location. These conflicts of use and pollution can be felt especially if the use

Biosphere Reserve	Representative Area	Representative Ecosystems	Characteristic Flora	Characteristic Fauna
South-East Rügen	Mecklenburg-Vorpommern coastal area	Beech forests, oligotrophic grassland and semi-arid grassland, eroding and balancing coasts, salt meadows	Wood Anemone (*Anemone nemorosa*), Everlasting Flower (*Helichrysum arenarium*), Germander Speedwell (*Veronica teucrium*), Cowslip (*Primula veris*), Sea Holly (*Eryngium maritimum*)	Sand Martin (*Riparia riparia*), Geese (*Anatidae* spp.), Atlantic Herring (*Clupea harengus*)
Schleswig-Holstein, Hamburg and Lower Saxon Wadden Sea	Mudflats, islands and marshes	Mudflats, salt marshes, sand dunes, dune islands	Glasswort (*Salicornia* spp.), Sea-Aster (*Aster tripolium*), Sea Buckthorn (*Hippophae rhamnoides*), Common Crowberry (*Empetrum nigrum*), micro and macroalgae, i.e. diatomes	Arctic Wading Birds (*Limicolae* spp.), Geese (*Anserinae* spp.), ducks (*Anatinae* spp.), terns (*Sternidae* spp.), gulls (*Laridae* spp.), Common Seal (*Phoca vitulina*), Harbour Porpoise (*Phocoena phocoena*), Flat Fish (*plaice, Flounder, Sole - Pleuronectidae* spp., *Bothidae* spp., *Soleidae* spp.), Brown Shrimp (*Crangon crangon*)
Schaalsee	Baltic beech wood area within the biogeographical province of the Central and Eastern European forests	Beech forest, calcium-rich lakes and bogs, oat grass meadows	Beech (*Fagus sylvatica*), Saw-Sedge (*Cladium mariscus*), muskgrasses (*Characeae* spp.), orchids (*Orchidaceae* spp.), Tall Cottongrass (*Eriophorum angustifolium*), Wood Anemone (*Anemone nemorosa*)	Eurasian Otter (*Lutra lutra*), White-Tailed Eagle (*Haliaeetus albicilla*), Common Crane (*Grus grus*), geese (*Anserinae* spp.), Fire-Bellied Toad (*Bombina bombina*), Laveret (*Coregonus lavaretus*)
Schorfheide-Chorin	Northern German new moraine landscape	Beech and pine forests, fields, waterbodies and moors	Adders' Tongue (*Ophioglossum vulgatum*), Wild Rosemary (*Ledum palustre*), Bogbean (*Menianthes trifoliata*), Everlasting Flower (*Helichrysum arenaria*)	Eurasian Beaver (*Castor fiber albicus*), Eurasian Otter (*Lutra lutra*), Common Crane (*Grus grus*), Lesser-Spotted Eagle (*Aquila pomarina*), Osprey (*Pandion haliaetus*), White-Tailed Eagle (*Haliaeetus albicilla*), European Pond Tortoise (*Emys orbicularis*)
Elbe River Landscape	Plains and glacial valleys	River, softwood alluvial woodland, hardwood alluvial woodland, alluvial grassland, dead arms, inland dunes	Pendunculate Oak (*Quercus robur*), Common Pear (*Pyrus pyraster*), Crab Apple (*Malus sylvestris*), Black Poplar (*Populus nigra*), Siberian Iris (*Iris sibirica*), Floating Moss (*Salvinia natans*), Water Chestnut (*Trapa natans*), *Jurinea cyanoides*	Eurasian Beaver (*Castor fiber albicus*), Red Kite (*Milvus milvus*), White Stork (*Ciconia ciconia*), Geese (*Anserinae* spp.), Middle Spotted Woodpecker (*Dendrocopos medius*), River Lamprey (*Lampetra fluviatilis*), Fire-Bellied Toad (*Bombina bombina*), Longhorn Beetle (*Cerambyx cerdo*), *Gomphus flavipes*
Spree Forest	Northern German old moraine landscape	Intermittent and alluvial woodland, wetland meadows, flowing waters	Slender Spiked Sedge (*Carex gracilis*), Meadow Fleabane (*Inula britannica*), Slenderstem Peavine (*Lathyrus palustris*), Marsh Marigold (*Caltha palustris*), Water Soldier (*Stratiotes alloides*)	Eurasian Otter (*Lutra lutra*), Black Stork (*Ciconia nigra*), White Stork (*Ciconia ciconia*), Common Crane (*Grus grus*), Woodpeckers (*Dendrocopos* spp., *Dryocopus* spp.), Burbot (*Lota lota*), Dragonflies (*Odonata* spp.)
Upper Lausitz Heath and Pond Landscape	Upper Lausitz heath and pond area	Ponds, moors, heaths, pine forests, alluvial meadows	White Water Lily (*Nymphaea alba*), Cranberry (*Oxycoccus palustris*), Bog Heather (*Erica tetralix*), Wild Rosemary (*Ledum palustre*), European White Elm (*Ulmus laevis*)	Eurasian Otter (*Lutra lutra*), Common Crane (*Grus grus*), European Nightjar (*Caprimulgus europaeus*), White-Tailed Eagle (*Haliaeetus albicilla*), Black Woodpecker (*Dryocopus martius*), Adder (*Vipera berus*)
Vessertal-Thuringian-Forest	Thuringian-Franconian medium-range mountains	Mountain spruce forests, mixed alpine forest (beech-dominated), alpine meadows, moors, mountain streams	Beech (*Fagus silvatica*), Silver Fir (*Abies alba*), Arnica (*Arnica montana*), Brookline (*Veronica beccabunga*), Common Alder (*Alnus glutinosa*), Cranberry (*Oxycoccus palustris*)	Black Woodpecker (*Dryocopus martius*), Woodcock (*Scolopax rusticola*), Red Deer (*Cervus elaphus*), Bog Bush-Cricket (*Metrioptera brachyptera*), European Fire Salamander (*Salamandra salamandra*), Dipper (*Cinclus cinclus*), Brown Trout (*Salmo trutta f. fario*)
Rhön	Central German mountainous land	Beech forests, lime-maple gorge forests, alpine meadows, oligotrophic and semi-arid grassland, basalt block heaps, moors	Beech (*Fagus sylvatica*), Stemless Carline Thistle (*Carlina acaulis* ssp. *caulescens*), Matgrass (*Nardus strictus*), Yellow Oat Grass (*Trisetum flavescens*), Arnica (*Arnica montana*), Globeflower (*Trollius europäus*), orchids (*Orchidaceae* spp.)	Black Grouse (*Tetrao tetrix*), European Eagle Owl (*Bubo bubo*), Black Stork (*Ciconia nigra*), Whinchat (*Saxicola rubetra*), Meadow Pipit (*Anthus pratensis*), Black Woodpecker (*Dryocopus martius*), Woodchat Shrike (*Lanius senator*), Red Kite (*Milvus milvus*), Corncrake (*Crex crex*), Common Snipe (*Gallinago gallinago*), Hermit (*Chazara briseis*), *Bithynella compressa*
Palatinate Forest-North Vosges (only German part)	South-west German stratified land	Beech and pine forests, grape-growing country, dystrophic waterbodies	Sweet Chestnut (*Castanea sativa*), Pasque Flower (*Pulsatilla vulgaris*), Wild Tulip (*Tulipa sylvestris*)	Wild Cat (*Felis sylvestris*), Black Woodpecker (*Dryocopus martius*), Peregrine Falcon (*Falco peregrinus*), Rock Bunting (*Emberiza cia*)
Bavarian Forest	Upper Palatinate and Bavarian Forest	Mixed alpine forests, pine forests, upland moors	Spruce (*Picea abies*), Silver Fir (*Abies alba*), Beech (*Fagus silvatica*), Snowbell (*Soldanella montana*), Chickweed Wintergreen (*Trientalis europea*)	Lynx (*Lynx lynx*), Eurasian Otter (*Lutra lutra*), Pygmy Owl (*Glaucidium passerinum*), White-Backed Woodpecker (*Dendrocopos leucotos*), Capercaillie (*Tetrao urogallus*)
Berchtesgaden	Northern chalk Alps	Mixed alpine forests, sub-alpine forests, chalk alpine meadows, oligotrophic lakes	(*Lomatogonium carinthiacum*), Aquilegia einseliana, Gentian (*Gentianaceae* spp.), Edelweiss (*Leontopodium alpinum*)	Alpine Marmot (*Marmota marmota*), Chamois (*Rupicapra rupicapra*), Golden Eagle (*Aquila chysaetos*), Ptarmigan (*Lagopus mutus*), Migrating Brown Trout (*Salmo trutta f. lacustris*)

Tab. 2: Biotic resources of the biosphere reserves (after AGBR 1995: 17; supplements: information from the biosphere reserves 2003)

is not sustainable and is not compatible with the goals of the biosphere reserve in question. It is therefore a major task for the BR administrations to solve these conflicts and to put sustainable economic development to the fore in further development (cf Chapter 3.5).
Under the current general economic and socio-political conditions, the cultural landscapes in a few biosphere reserves are faced with a long-term far-reaching change. There is a risk that changing or surrendering the land use will alter the landscape as a basis for recreational use to its detriment and will jeopardise the conservation of the species and habitats that depend on the continued existence of these ecosystems of the cultural landscape. Cultural landscapes depend on continuous use for their long-term conservation. Use can only be expected if landscape maintenance is recognised and encouraged as a task for society (MAYERL, D. 1990; cf Chapter 3.2.4).

Cultural landscape in the Schorfheide-Chorin Biosphere Reserve

Tasks and Management

Germany is a federal state, the Federal Government having the framework legislative competence for nature conservation and landscape management. However, the conservation, maintenance and development of the individual biosphere reserve are in the responsibility of the *Länder*, including the legal safeguarding of the biosphere reserves, the establishment of the administration and the implementation of the Guidelines for Conservation, Maintenance and Development (AGBR 1995).
The BR administrations have to perform diverse administrative and technical tasks. For this, the competent *Land* authorities give them sufficient relevant powers. Depending on the individual regulation, the administrative tasks of the administrations can range from statements by the authorities to participation in the planning and approval procedure right up to performing sovereign tasks independently.

The wide-ranging tasks in the UNESCO biosphere reserves require matching staff and financial resources. The management spectrum in the areas and the intended external effect as model regions for sustainable development must be reflected in expert staff in the administrative authorities. An effective administration is a prerequisite for a successful management.
In their work the administrations can be supported by various bodies and institutions. Both advisory councils and boards are mainly made up of representatives of the various user groups and associations as well as representatives of local and *Land* politics or of the administrations in question. Both provide technical advice to the BR administration and harmonise measures with the concerns of the local communities and local or technical authorities and associations concerned.
Cooperation with the relevant groups and institutions in society must have high priority so that the action of the administrations leads to a positive response and high levels of acceptance among the public.
In line with the statutes, foundations and sponsoring bodies support the administration in carrying out its tasks, both in terms of ideas and material, for example by means of targeted public relations work or by recruiting donors.
But the administrations of biosphere reserves can also contractually entrust them with clearly defined tasks, for example environmental education (as in the Bavarian part of the Rhön BR) or area management in return for reimbursement of costs. The establishment of dedicated companies, as practised for area management in the Schleswig-Holstein Wadden Sea BR (NationalparkService GmbH), is suitable for this.

Geographic Breakdown

The goals and tasks in biosphere reserves need to be broken down geographically. Zones with different functions and areas of responsibility are specified according to the impact of human activity: **core area, buffer zone** and **transition area**. The latter may also contain a regeneration zone (ERDMANN, K.-H., NAUBER, J. 1991).
With regard to the biosphere reserves' task of being model regions for sustainable economic development, the transition area should be large enough – usually more than half of the area of the entire reserve (GERMAN MAB NATIONAL COMMITTEE 1996; Criterion 7).
In an intensive discussion process the Permanent Working Group of the Biosphere Reserves in Germany (AGBR 1995: 5) agreed on a nationwide definition of biosphere reserves (see above). The following has also been defined for the geographic breakdown of the zones in Germany:

Zones in a Biosphere Reserve, German Definitions

Core area
Every biosphere has a core area where nature can develop without being influenced by people as far as possible. The aim is to exclude human use from the core area. The core area should be large enough to enable the dynamism of ecosystemic processes. It can comprise several partial areas. The conservation of natural or semi-natural ecosystems has top priority. Research activities and surveys on ecological environmental observation must avoid disturbance to ecosystems.
The core area must be legally protected as a national park or a nature conservation area.

Buffer zone
The buffer zone is used to conserve and maintain ecosystems created or influenced by human use. The buffer zone should screen the core area against damage. The principal aim is to conserve cultural landscapes comprising a broad spectrum of different habitats for a large number of animal and plant species – including endangered species – typical of the living space. This should be achieved mainly by landscape management. Recreation and measures for environmental education should be geared towards the purpose of conservation. In the buffer zone the structure and functioning of ecosystems and the natural balance are examined and ecological environmental monitoring is conducted.
The buffer zone should be legally protected as a national park or a nature conservation area. If this has not yet been done, protection of this kind must be aimed at. The conservation status of conservation areas that have already been designated must not be reduced.

Transition area
The transition area is the area where the human population lives, works and rests. The aim is to develop a way of working that does equal justice to the needs of man and nature. Socially compatible production and marketing of environmentally friendly products contribute to sustainable development. Sustainable uses in particular characterise the landscape in the transition area. This is where there are opportunities for developing environmentally friendly and socially compatible tourism.
Primarily man-environment relations are researched in the transition area. At the same time, the structure and functioning of ecosystems and the natural balance are investigated and ecological environmental monitoring and measures for environmental education are conducted. Seriously impaired areas can be adopted as regeneration zones within the transition area. In these areas the emphasis of the measures is on remedying damage to the landscape.
Areas worthy of protection in terms of nature conservation within the transition area should be legally protected by means of designation as a conservation area and by town and country planning instruments.

(AGBR 1995: 12-13)

It is specifically the definition of the transition area that makes clear the task of sustainable economic development in the biosphere reserves. This will have to be strengthened in the future (cf Chapter 3.5).

Legal Safeguarding of the Biosphere Reserves

The statutory regulation for biosphere reserves in the Federal Republic and in the *Länder* is still neither uniform nor satisfactory. The biosphere reserves that the Unification Treaty took over from the GDR National Park Programme of 1990 with their uniform regulations have legal force.
The eastern German *Länder* understood biosphere reserves as conservation areas with an overall regulation, but expressly with a commission to develop that goes beyond the conservation function. The western German *Länder* mainly saw biosphere reserves as a regional planning instrument. On this understanding, biosphere reserves are planning and development areas – without an overall regulation – in which areas worthy of protection are protected with the existing nature conservation legislative instruments.
In the years from 1995, an intensive discussion about the framework legislative regulation of biosphere reserves was conducted between the Federal Government and the *Länder.* With the 3rd amendment of the Federal Nature Conservation Act of 26 August 1998 the Federal Government introduced a regulation for protected areas (without a reference to UNESCO) that opens up the opportunity for the *Länder* to deviate from it in their *Land* Nature Conservation Acts. This means that UNESCO biosphere reserves can also be legally designed as planning and development areas.

In the current framework act, the Federal Nature Conservation Act of 25 March 2002, the provision for biosphere reserves is worded as follows:

Extract from the Federal Nature Conservation Act
(25 March 2002)

Article 25 Biosphere Reserves:
(1) *Biosphere reserves are designated, legally binding areas enjoying uniform levels of protection and development, and which:*
1. are large and are typical representatives of certain landscape types
2. fulfil the requirements for nature conservation areas in the greater part of their territory, and most of the requirements for landscape reserves throughout the rest of their territory
3. serve the primary purpose of preserving, developing or restoring landscapes shaped by traditional, diverse forms of use, along with their species and biotope diversity as evolved over time, including wild forms and formerly cultivated forms of commercially used or usable animal and plant species, and
4. illustrate ways of developing and testing forms of economic activity that are especially conserving of natural resources.
(2) *The Federal Länder shall ensure that biosphere reserves are developed with due regard for the exceptions concerning core areas, buffer zones and transition areas required as a result of biospheres' large size and inclusion of populated areas, and receive the same level of protection afforded to nature conservation reserves or landscape reserves.*

Article 22 Designation as a Protected Area
(1) - (3)
(4) *The Federal Länder may adopt diverging regulations for biosphere reserves and nature parks. (...)*
(Bundesgesetzblatt I S. 1193)

Overall, this framework legislative regulation by the Federal Government cannot be satisfactory, either in terms of content or from the point of view of the MAB Programme. On the one hand, there is no clear purpose for biosphere reserves as model regions for sustainable development, on the other hand it is not UNESCO biosphere reserves that are legally enshrined, but biosphere reserves without UNESCO recognition.
This means that the Federal legislative regulation does not satisfy the international requirements and UNESCO programmes, such as the Seville Strategy and the International Guidelines for Biosphere Reserves (UNESCO 1996). Many *Länder* had already laid down biosphere reserves in their *Land* Nature Conservation Acts before this framework legislative regulation. This results in differing regulations. Biosphere reserves in the eastern German *Länder* are now no longer consistently specified with a single regulation. For example, in the *Land* Brandenburg under Article 25 of the Nature Conservation Act of 25 June 1992, biosphere reserves are declared by promulgation by the supreme nature conservation authority. One example from a *Land* regulation that consistently follows the UNESCO requirements for the MAB Programme is the Bavarian Nature Conservation Act:

Extract from the Bavarian Nature Conservation Act of 18 August 1998:

Art. 3a: Biosphere Reserves (in Section II "Landscape Planning and Landscape Management")
(1) The State Ministry for Land Development and Environmental Issues may declare large-scale, representative sections of cultural landscapes as biosphere reserves after recognition by the United Nations Organisation for Education, Science and Culture. The aim of biosphere reserves is to exemplify
1. the protection, conservation and development of cultural landscapes,
2. the development of a sustainable way of working that does equal justice to the needs of man and nature,
3. environmental education, ecological environmental monitoring and research.
(2) Biosphere reserves shall be broken down into core areas, buffer zones and transition areas according to the influence of human activity.
(3) The term Biosphere Reserve may be used only for the areas declared pursuant to (1).
(Gesetz- und Verordnungsblatt (Bayern) 1998: 59, non-official translation)

Cooperation and Exchange of Experience within the Network of German Biosphere Reserves

Even immediately after German Unification and due to the biosphere reserves that were taken over in the Unification Treaty via the National Park Programme of the former GDR it became clear that a nationwide working body is urgently needed for harmonisation and for the exchange of experience. Thus, in 1991 the German MAB National Committee and the *Länder* involved through the recognised biosphere reserves set

up the Permanent Working Group of the Biosphere Reserves in Germany (AGBR). It is made up of the heads of the BR administrations and representatives from the Federal Government and the *Länder*. The AGBR started work immediately. From 1996 onwards the Working Group sought more contact and cooperation with the environmental associations, relevant foundations and business associations. This opening up to these relevant groups in society extended the basis for the work of the Working Group.
In the 13 years of its existence, the Working Group has established itself as an effective body for a harmonised development of the UNESCO biosphere reserves in Germany. In this time it has passed some key milestones:
- Guidelines for the conservation, maintenance and development of biosphere reserves in Germany (AGBR 1995) as a pioneering instrument for the further development of the UNESCO biosphere reserves,
- participation in the development of the "Criteria for Designation and Evaluation of UNESCO Biosphere Reserves in Germany" of the German MAB National Committee (National Criteria cf Annex, p. xxx),
- elaboration of general guiding principles for biosphere reserves in Germany, adopted at the 22nd meeting of the AGBR in the South-East Rügen BR on 29 September 1999,
- "Biosphere Reserves in Germany: It's Worth a Trip – And So Is Staying"; published by EUROPARC Deutschland together with the Working Group (2002) on environmentally friendly tourism in the biosphere reserves.

Conclusion on the Current State of Development

Previous efforts, successes and milestones in the development of the network of biosphere reserves must not conceal the fact that a great deal remains to be done for the acceptance, importance and the image of the biosphere reserves in Germany. The public perception of biosphere reserves still leaves a lot to be desired. Often enough, biosphere reserves are equated with classic conservation areas, such as national parks. From this one can conclude that the contribution that biosphere reserves make to conservation and maintenance is placed to much to the fore and that the task as a model region for sustainable development is not publicised.
The tension between conservation and development must be solved in the plans for biosphere reserves in close cooperation with the affected parties. In this connection, the weighting must be shifted in favour of sustainable regional development and strengthening the image and the recognition factor of the biosphere reserve. Action in biosphere reserves should be more geared towards work in Agenda 21.
The designation of the areas as "UNESCO biosphere reserve" has had a very positive impact in applications for project and promotional funding in national and European competition with other regions, e.g. LEADER, INTEREG or from *Land* funding programmes.
On the whole, it can be concluded that a great deal has been achieved, but that future development will have to be even more geared towards sustainable economic development (cf Chapter 3.5).

Literature

AGBR (Ständige Arbeitsgruppe der Biosphärenreservate in Deutschland) (1995): Biosphärenreservate in Deutschland. Leitlinien für Schutz, Pflege und Entwicklung, Berlin, Heidelberg, etc.
BfN (Bundesamt für Naturschutz) (2002): Daten zur Natur, Bonn.
Bundesgesetzblatt 2002 Teil I Nr. 22 vom 3. April 2002: Gesetz zur Neuregelung des Rechts des Naturschutzes und der Landschaftspflege und zur Anpassung anderer Rechtsvorschriften vom 25. März 2002: p. 1193 - 1218; § 22, 25: 1202)
Deutsches MAB-Nationalkomitee (1995): Empfehlung für den Beitrag Deutschlands zum UNESCO-Programm "Der Mensch und die Biosphäre"(MAB) für den Zeitraum 1996 bis 2001 (vierter mittelfristiger Plan der UNESCO), Bonn.
Erdmann, K.-H. (1997): Biosphärenreservate der UNESCO: Schutz der Natur durch eine dauerhaftumweltgerechte Entwicklung. In: Erdmann, K.-H., Spandau, L. (Ed.): Naturschutz in Deutschland. Strategien, Lösungen, Perspektiven, Stuttgart.
Erdmann, K.-H., Nauber, J. (1991): UNESCO-Biosphärenreservate. Ein internationales Programm zum Schutz, zur Pflege und zur Entwicklung von Natur- und Kulturlandschaften. In: Umwelt. Informationen des Bundesministers für Umwelt, Naturschutz und Reaktorsicherheit 10/1991, pp. 440-450.
Erdmann, K.-H., Nauber, J. (1995): Der deutsche Beitrag zum UNESCO-Programm "Der Mensch und die Biosphäre" (MAB) from July 1992 to June 1994. With an English Summary, Bonn.
EUROPARC Deutschland (Ed.) (2002): Biosphere Reserves in Germany: It's Worth a Trip And So Is Staying, Berlin.
Gesetz- und Verordnungsblatt (Bayern): Gesetz über den Schutz der Natur, die Pflege der Landschaft und die Erholung in der freien Natur (Bayerisches Naturschutzgesetz) in der Fassung der Bekanntmachung vom 18. August 1998, p. 593.
Gesetz- und Verordnungsblatt (Brandenburg): Gesetz über den Naturschutz und die Landschaftspflege im *Land* Brandenburg (Brandenburgisches Naturschutzgesetz) vom 25. Juni 1992; I/92, p. 208
German MAB National Committee (Ed.) (1996): Criteria for Designation and Evaluation of UNESCO Biosphere Reserves in Germany, Bonn.
Mayerl, D. (1990): Die Landschaftspflege im Spannungsfeld zwischen gezieltem Eingreifen und natürlicher Entwicklung – Standort und Zielsetzung, Planung und Umsetzung in Bayern. In: Natur und Landschaft 65(4), pp. 167-175.
UNESCO (Ed.) (1996): Biosphere Reserves. The Seville Strategy and the Statutory Framework of the World Network, Paris.

New Concepts for the Model Regions

3.

3.1 *Man and the Biosphere*

3.1.1 *People and Cultures in Biosphere Reserves*

Lenelis Kruse-Graumann

People in Biosphere Reserves: Where Are They?

The MAB Programme: theory and reality

The MAB Programme and the concept of biosphere reserves were already signed up to the idea of sustainable development in the 1970s, even before it became the official guiding principle for the 21st century. However, a great deal remains to be done to do justice to the demands associated with it. The impetus from the Rio Conference (1992) and the development of the Seville Strategy for Biosphere Reserves (1995) have opened up new perspectives and set new emphases.

To implement the balance between preserving the natural foundations and people's demands for living and survival with their various historically, culturally and socially shaped identities, more attention must be paid to the people in the biosphere reserves and the surrounding areas. People with their various group memberships, roles, interests and responsibilities must repeatedly be given the opportunity to "appropriate" these areas for themselves again, to preserve them and to develop them along sustainable lines.

Culture is the way and means for people to interact with nature

Appropriation is used to refer to people's activities to define, and/or change the nature surrounding them, i.e. the human environment, for their own purposes. It is only in this way that it becomes "their" environment.

On the one hand, appropriation is achieved by changing nature by means of work: creating rice terraces, removing stones to build houses, developing transport routes on water and land, breeding plants as well as appropriation by means of profit-oriented exploitation and consumption of vital resources, by means of the subjugation of other people and peoples.

On the other hand, there is also the **symbolic** appropriation of nature and/or its cultivation, for example by means of linguistic terms ("weeds", "buffer zone"), artistic representations, myths and stories. Knowledge and experience, as well as religion and tradition play a key role in this. The value that depends on culture-specifics is closely linked to symbolic appropriation. Attitudes and practices can be evaluated so differently that understanding between cultures is often difficult, if not impossible.

Appropriation is a continuous process: as people appropriate their environment (e.g. by working and using it), they change the nature surrounding it to the same extent. In turn, this changed environment has an impact on humans, who then change again. This interaction constantly repeats itself.

During appropriation, two distinct, but associated, processes take place: on the one hand, the human race (with its peoples and tribes, their cultures and languages) appropriates nature (its raw materials and forces) over many generations and leads to very different philosophies and practices in different cultural groups. On the other hand, however, every single person appropriates his or her environment afresh because he or she has to learn the achievements of his or her culture in his or her life from scratch.

How, for example, does a person appropriate a landscape such as that found in a biosphere? Every new child, as well as every tourist or new inhabitant, appropriates what previous generations have created, for example by exploring the areas, by moving through the landscape, by using things, but also by naming and appreciating things that need to be protected or husbanded or that can be marketed or have to be conserved for future generations as (natural or cultural) "heritage". In the most favourable case, appropriation goes hand in hand with a positive identification with the place, expressed, for example, in sentences such as "I am a Spreewälder". On the other hand, various individuals or interest groups can have very different ideas about appropriation and corresponding intentions for action that are not infrequently mutually exclusive and can lead to conflicts (cf Graumann, C. F. 1996).

People living in a biosphere reserve: Family party in Friedersdorf as part of the environmental education programme in the Upper Lausitz Heath and Pond Biosphere Reserve (cf Chapter 4.8)

Nature conservation through cultural conservation or vice versa?

If there is currently renewed interest in the so-called indigenous knowledge of the people living in approximately 5,000 tribal communities (a total of around 200-300 million, half of them in China and India), it is because it has been recognised that many indigenous or, more broadly interpreted, traditional societies have succeeded in living with nature compatibly or sustainably.

To do this, they use practices supported by a complex interaction of knowledge, philosophies and religious belief systems that have come about over long periods of time. "Such people walk lightly on the landscape" say McNeely and Keeton (1995), and a large number of careful studies of indigenous and local cultures has now demonstrated how this works (e.g. UNEP 1999; UNESCO 2002). They show an impressive networking of nature and culture, where culture not only means careful dealings with the landscape, plants and animals, but also how community members deal with each other. Traditional ecological knowledge (TEK) is thus the component of all of these cultures that gives them their identities. However, the studies also bear witness to failed adaptations and appropriations, of conflicting developments and collapses of cultures (e.g. Easter Island) that often come about due to the intervention of the industrialised, western world. Now, various conditions are known under which the knowledge systems and cultural forms of dealing with natural resources and fellow humans that are so well adapted to local circumstances can be maintained. They include, for example, moderate group sizes that do not exceed the carrying capacity of a landscape, good cohesion in the groups and full control of local resources. However, these cultural patterns have proved to be extremely susceptible and ever more at risk in an increasingly globalised world.

Above and beyond the indigenous peoples, there are other local communities who adapt their behaviour, their language, their social structures to changing living conditions in historical continuity with countless adaptation processes and have thus created the cultural diversity with which we are familiar today, for example in many biosphere reserves. The diverse knowledge systems and ways that the local people have of dealing with the natural environment can be seen in biosphere reserves. They have thus become cultural landscapes in the best sense of the words (cf Kruse-Graumann, L. 2002).

In a world determined by technology, which has also been dubbed the "second" nature for humans, but today has become the "first" nature for most people, the indigenous patterns of nature appropriation have largely disappeared. Emancipation from natural living conditions has long become alienation. Mastery of nature, exploitation and destruction are the modern appropriation patterns, the consequences of which are ever more obvious.

The fact that we need to think about our relationship with nature more than ever is a characteristic of our present culture. In the 21st century we have to find the way towards sustainable development and this presupposes that we will develop an awareness of the mutual dependency of nature and culture, of biological and cultural diversity. Nature conservation and, in the broader sense, sustainable development are thus inconceivable without cultural change.

A successful policy for sustainable development must acknowledge that:

- The conservation of nature and its biological diversity is in the interest of humans because they are natural and cultural beings and cultural diversity builds upon biological diversity.
- The protection of cultural diversity is required because humans must be involved as cultural beings as part of the biosphere, but not reduced to their biological parts.
- Moreover, the protection of cultural diversity is a guarantee for the conservation of biological diversity, especially in the regions where local and indigenous communities have successfully conserved this diversity (cf Kruse, C. F. 2001).

Walking and sports, as in the Berchtesgaden Biosphere Reserve, are examples for the appropriation of landscape in biosphere reserves.

People: Their Roles in Biosphere Reserves

If we leaf through the so-called green book on "Biosphere Reserves in Germany" (1995), we find lots of pretty photographs, but only a few of people or human-made constructions, such as churches, houses or vineyards. This barely illustrates the diversity of interactions between people and the landscape that turns a biosphere reserve into what is protected, husbanded and further developed today.

The UNESCO book "Biosphere Reserves. Special Places for People and Nature" (2002) is very different, where people in their various social roles and cultural identities and their links with nature and the cultural environment are given disproportionately more – visual – prominence.

And finally, a critical look at the Seville Strategy, the binding management philosophy and the catalogue of targets for biosphere reserves: it states that biosphere reserves can only be examples for sustainable development if they meet all of the cultural, spiritual and economic needs of society. The call to emphasise the human dimension of biosphere reserves is very general and the objective of creating a relationship between cultural and biological diversity and preserving traditional knowledge is very ambitious.

A general theme can be found in all Seville documents: the incorporation of the demands of various stakeholders and the comprehensive participation of these groups in planning and decision-making processes.

This prescribes a comprehensive catalogue of tasks; because it is ultimately expected

- that the various conditions (e.g. modalities of appropriation) and their links that determine the interactions between people and the biosphere reserve, are familiar,
- that their importance for various tasks (e.g. strengthening nature conservation or encouraging certain economic activities) can be assessed,
- that it is known what should be conserved and strengthened (e.g. traditional practices of land and landscape management),
- that it is known what perception and action patterns (e.g. images of nature, lifestyles, consumption habits and economic practices) can act as obstacles or as supports in the necessary process of a social change towards sustainability and
- that all of this is implemented in a biosphere reserve.

In recent years (especially in Germany) numerous efforts (conferences, workshops, projects) to address such issues in connection with nature conservation and utilisation in large-scale protected areas from very different social and behavioural science perspectives are remarkable (e.g. Erdmann, K. H., Schell, C. 2002; Kruse-Graumann, L. et al. 1995; Schweppe-Kraft, B. 2000; Grewer, A. et al. 2000 and O'Riordan, T., Stoll-Kleemann, S. 2002).

For example, just one important issue for the management of biosphere reserves is presented: how can **communication and participation** in biosphere reserves be developed, what conditions need to be considered?

Often, research on these issues (still) goes under the buzzword of "acceptance" (e.g. Hofinger, G. 2001; Schuster, K. 2003; Stoll, S. 1999). This is easy to explain as in the past it was mainly a question of designating (nature) conservation areas and the problem of whether and how the local population could be induced to accept the "top down" conditions and restrictions.

How do acceptance problems manifest themselves, how can they be explained and what measures can be taken to mitigate them or to secure the willingness of the local population to participate and cooperate? Below, you will find just one study that demonstrates interesting approaches and notable conclusions:

In the "Schorfheide Project" Hofinger dealt with acceptance problems in the Biosphere Reserve (HOFINGER, G. 2001 a and b; cf DÖRNER, D. et al. 1995). On the basis of interview data collected by "key people" in the Biosphere Reserve five times within two and a half years, she succeeded in analysing the ideas associated with the Biosphere Reserve, how they are evaluated and how acceptance or rejection of the Biosphere Reserve change over time – also depending on the various perceptions.

Hofinger found very different attitudes to the Biosphere Reserve among her interviewees: for example, it is seen as

- *a region or landscape;*
- *a protected area with UNESCO recognition;*
- *an instrument of species and nature conservation or as the embodiment of nature conservation;*
- *an instrument of regional development or also as*
- *an authority and its employees.*

Whereas the first three attitudes are mostly associated with positive evaluations, the "authority" is usually evaluated critically to negatively. Rejection also predominates with the "nature conservation instrument", but for quite differing reasons: for some people nature conservation goes too far, for others it does not go far enough. A more detailed analysis of the responses to a very simple question was revealing: "If there were a referendum now – seven years after establishment of the Biosphere Reserve – how would you vote? Should the Biosphere Reserve continue to exist?" If 68 per cent of those questioned answered with "yes", 17 per cent with "yes, but", 13 per cent with "no" and two per cent with "no, but", this result would satisfy all those who are only interested in numbers. But it would not add much to our findings if we did not look into the reasons for these decisions. Thus, among those who said "yes" and those who said "no" (as well as the undecided in both direction), there were once again various reasons for their responses: the reasons cited by advocates include "nature conservation, forest conservation, preservation of livelihoods", but also "protection against exploitation by 'Wessis'" [people from the former West Germany] or "hope for benefits for the region". The "abolitionists'" comments include "criticism of the administration" or "hampering of their own goals".

Hofinger went even further with her analyses and described seven different forms of acceptance, which she called "active opposition", "rejection", "tolerance", "indifference", "approval", "enthusiasm" and "ambivalence" and which include evaluations, emotional references and tendencies for action. The most common forms of acceptance that she found were tolerance, indifference and approval. At the end of the 1990s there was no openly waged hostility in the Schorfheide-Chorin Biosphere Reserve. Further distinctions were revealed when the forms of acceptance were analysed in more detail for different professional groups in the Biosphere Reserve. Thus, for example, approval of conservation for the region was found most among farmers, associated with tolerance for the authority. Approval was most common "as long as there is money, advice and support". The majority of those involved in tourism agree because they see the conservation of nature in the Biosphere Reserve as their livelihood. By contrast, the forestry employees already see themselves as nature conservationists, but reject the type of "biosphere reserve" with the totally protected core areas.

Humans are moving towards the centre of interest: Biosphere reserves aim at balancing their different needs.

This study shows how different the ideas, especially the associated evaluations of biosphere reserves among the members of various professional groups or stakeholders, can be. Above all, these and similar results on different concepts and

images of "nature" or "natural beauty" or "nature conservation" clearly illustrate that communication processes and participation projects, and moreover intervention projects to change behaviour, are planned for specific target groups and "must address people where they happen to be".
Here, only one example from a growing number of research projects could be quoted. On the whole, they illustrate that more applied social science research projects are needed in the biosphere reserves. Only in this way can the potential determining factors (both in the interest of conserving tried and tested belief systems and behaviour patterns, as well as in the interest of changing detrimental ones) be tested for their impact and importance for participation and appropriation processes.

Conclusions

If people as members of social and ethnic groups, of family and professional groups, with their (traditional) knowledge and belief systems, their attitudes and values, their prejudices and oppositions have played only a minor role in the MAB Programme and the management practice of biosphere reserves for a long time, the adoption of the guiding principle of sustainable development proclaimed for the 21st century means a new challenge for the biosphere reserves. Good examples of sustainable development in biosphere reserves should disseminate globally. If they will be able to do justice to this ambitious goal, remains to be seen. One thing can be stated though: People ought to be moved to the centre of interest.
The involvement of interest groups, the active participation of the local population and other relevant groups in the planning and design of sustainable development in biosphere reserves require that we know something about the conditions that motivate the people to accept and promote the need to protect areas, to use resources sparingly, to develop sustainably produced and, at the same time, attractive products, or to oppose these objectives. Only knowledge of this kind helps to advance the further development of the biosphere reserves as model regions for sustainable development, to overcome conflicts and to communicate the importance of biosphere reserves and their various functions internally and externally. For this, technical skills are needed that are not usually available in the biosphere reserves. More than in the past, economic, social and behavioural science research projects must be located in biosphere reserves to provide the necessary findings, to help to develop management and communication strategies and to support social and societal monitoring. But, research-based communication, participation and teaching methods must be deployed professionally. The moderation of participation processes or even mediation in conflicts in biosphere reserves requires experience and skills that go far beyond nature conservation training. The involvement of people in the interest of cooperative administration and design of conservation and utilisation strategies therefore not only requires the incorporation of nature conservation experts, but also economics and tourism experts, communication advisors and marketing specialists, who together have the task of constantly creating new "appropriation possibilities" for the local population, visitors and society as a whole.
Biosphere reserves as model regions for sustainable development? An ambitious goal that requires a precise analysis of the prerequisites and the will to provide the necessary human and monetary resources.

Literature

AGBR (Ständige Arbeitsgruppe der Biosphärenreservate in Deutschland) (1995): Biosphärenreservate in Deutschland, Berlin: Springer.
Dörner, D., Kruse, L., & E.D. Lantermann. (1995): Umweltbewusstsein, Umwelthandeln, Werte und Wertewandel. Zur Erforschung der Bedingungen und Formen anwendungsorientierten ökologischen Lernens. Begleituntersuchung der Etablierung des Biosphärenreservates Schorfheide-Chorin. In: K.-H., Erdmann, Nauber, J.: Der deutsche Beitrag zum UNESCO Programm "Der Mensch und die Biosphäre" (MAB), Bonn, pp. 73-96.
Erdmann, K.-H., Schell, C. (Eds.) (2002): Naturschutz und gesellschaftliches Handeln, Bonn: Bundesamt für Naturschutz.
Graumann, C. F. (1996): Aneignung. In: Kruse, L., Graumann, C. F., Lantermann, E. D.: Ökologische Psychologie, Weinheim: Beltz-PVU, pp. 124-130.
Grewer, A., Knödler-Bunte, E. Pape, K. & A. Vogel (Eds.) (2000): Umweltkommunikation. Öffentlichkeitsarbeit und Umweltbildung in Großschutzgebieten. Reihe: Luisenauer Gespräche. Bd. 1, Berlin: Kommunikation und Management Verlag.
Hofinger, G. (2001 a): Formen von "Akzeptanz" – Sichtweisen auf ein Biosphärenreservat. Umweltpsychologie, 5, pp. 10-27.
Hofinger, G. (2001 b): Denken über Umwelt und Natur. Weinheim: Beltz-PVU.
Kruse, L. (2001): Weltkultur und indigene Kulturen. In: Düssel, R., Edel, G. & U. Schödlbauer (Eds.): Die Macht der Differenzen, Heidelberg: Synchron, pp. 81-94.
Kruse-Graumann, L. (2002): Natur und Kultur – Vermächtnis und Zukunftsaufgabe. In: Erdmann, K.-H., Schell, C. (Eds.): Naturschutz und gesellschaftliches Handeln, Bonn: Bundesamt für Naturschutz, pp. 3-11.
Kruse-Graumann, L., v. Dewitz, F., Nauber, J. & A. Trimpin (1995): Societal Dimensions of Biosphere Reserves – Biosphere Reserves for People. Proceedings of the EuroMAB workshop, 23-25 January 1995, Königswinter. In: Deutsches MAB Nationalkomitee (Ed.): MAB Mitteilungen 41, Bonn.
McNeely A., Keeton, W. S. (1995): The Interaction between Biological and Cultural Diversity. In: van Droste, B., Plachter,

H. & M. Rössler (Eds.): Cultural Landscapes of Universal Value, Jena: Gustav Fischer Verlag, pp. 25-37.
O'Riordan, T., Stoll-Kleemann, S. (Eds.) (2002) Biodiversity, Sustainability and Human Communities. Protecting Beyond the Protected, Cambridge: Cambridge University Press.
Perez de Cuéllar, J. (1995): Our Creative Diversity. Report of the World Commission on Culture and Development, Paris: UNESCO.
Schuster, K. (2003): Lebensstil und Akzeptanz von Naturschutz, Heidelberg: Asanger.
Schweppe-Kraft, B. (2000): Innovativer Naturschutz – Partizipative und marktwirtschaftliche Instrumente, Bonn: Bundesamt für Naturschutz.
Stoll, S. (1999): Akzeptanzprobleme bei der Ausweisung von Großschutzgebieten, Frankfurt: Lang.
UNEP (1999): Cultural and Spiritual Values of Biodiversity. London, Intermediate Technology Publications.
UNESCO (Ed.) (1996): Biosphärenreservate. Die Sevilla Strategie und die Internationalen Leitlinien für das Weltnetz, Bonn: Bundesamt für Naturschutz.
UNESCO (Ed.) (2002): Biosphere Reserves: Special Places for People and Nature, Paris.

3.1.2 *From Environmental Education to Learning for Sustainability*

Gertrud Hein and Lenelis Kruse-Graumann

Environmental Education in Biosphere Reserves

From the outset, encouraging environmental education has been a central goal of the UNESCO Man and the Biosphere Programme (MAB), and thus a task for all biosphere reserves. Every biosphere reserve is required to develop and implement criteria and contents for environmental education in its framework plan, taking account of the specific structures of the biosphere reserve (German MAB National Committee 1996). There are thus diverse environmental education offers for the population and visitors in biosphere reserves: natural history excursions and seminars, nature experience programmes, project days, teaching and nature experience paths as well as information centres with exhibitions and extensive information offers. The most varied interest groups can obtain comprehensive information about the natural resources of the area, about the objectives and tasks of biosphere reserves and about the relationship between people and the environment. The educational offers meet with a very good response and are very important for the image and external impact of the biosphere reserves.

Education for Sustainable Development

With the Seville Strategy (1995) the MAB Programme extended its objectives: UNESCO biosphere reserves should become model regions for sustainable development. Nature and resource conservation as well as the preservation of ecosystems should now be seen in context and weighed up against human economic interests and equality of opportunity and/or equitable distribution for present and future generations. Education for sustainability thus becomes a wide-ranging and ambitious programme, where ecological, economic and socio-cultural aspects should be discussed, negotiated and, finally, implemented.

This means that many educational and learning processes must be initiated in every biosphere reserve for the people who live and work there, for children and young people who grow up in the biosphere reserve – and who want to earn their livelihoods there in the long run – for visitors and tourists who have maybe only seen "beautiful nature" in the past but have not really thought about how it came about, what role it may play above and beyond the local for the global situation of the earth system.

The successful further development of a biosphere reserve thus greatly depends on the extent to which the population identifies with the guiding principles of sustainable development and can be motivated to participate in shaping and implementing the biosphere reserve. Each individual has to become aware that he or she takes responsibility for the present and future generations and for the environment with every action as well as non-action (German MAB National Committee 1996).

People keep deciding for or against regional and sustainably produced products at the market, at the grocery or in restaurants.

People keep deciding for or against regional and sustainably produced products at the market, at the grocery or in restaurants. But the decision to buy and consume something specific, the choice of a mode of transport (bus or rail, their own car, bike) or the decision to build a low-energy house hardly depend on information alone (e.g. from brochures, presentations, press articles). Decisions of this kind are usually influenced by many other factors. We know that the balance of such decision processes is often not very environmentally friendly, maybe aimed at maximising profits in the short term individually, but is not very socially responsible. This is where changes from non-sustainable to sustainable development are needed: sustainable development presupposes a change to an array of culture and lifestyle specific action patterns and decision-making processes.

In biosphere reserves, in addition to the past educational offers, there are many means of using the example of specific projects to demonstrate and try out which ways lead to a new "sustainability awareness" and how people can be motivated to "sustainable" and "viable" behaviour. To extend traditional environmental education into "education for sustainable development", new concepts and new projects are needed in which the economic and socio-cultural requirements are also considered and weighted above and beyond the perspectives of the ecological dimension. Furthermore, education for sustainable development must be given a completely new standing within the functions of biosphere reserves: education and life-long learning are fundamental components of sustainable development, a process in which the global guiding principle of "sustainability" is constantly defined and realised by new local and regional (sub)objectives. An educational and learning process of this kind must be supported by integrated research approaches, in which not only natural and social sciences interact (ideally in an interdisciplinary way), but also in which the various groups of players in a biosphere reserve are continuously included to participate.

Learning for Sustainability: Prerequisites and Principles

It is well known that environmental problems are not problems of the environment, but problems of people in dealing with nature, resources and environmental pollution. This means that ultimately they are the consequence of maladapted, non-sustainable behaviour or action. The path to sustainability consequently means: changing and correcting behaviour and adapting it to new findings and circumstances. Specifically, this means: forgetting detrimental behaviour and learning more compatible, sustainable, viable behaviour (Kruse, L. 1999, 2002a and b).

The following must be remembered:

- Sustainable and non-sustainable behaviour patterns are not innate; they are learned and acquired from a young age and are constantly reinforced culturally and socially. Upbringing, education and learning are very important here, with as much attention having to be paid to relearning as to new learning.
- Action takes place at various levels of individual and collective action (individual, family, company, school, village community, local club, as well as region, nation and international community). Learning processes must therefore be shaped in as many different ways.
- Environmentally relevant and sustainable actions have special features that make learning and acquiring environmental and sustainability skills more difficult (cf Dörner, D. 1989). Actions can have direct impacts that can be perceived by all immediately. In the case of environmentally relevant, and thus sustainable, action, the causal link and the time and geographic effect can often not be perceived by the individual person – or only with a delay. Thus, it is hard for individuals to recognise what they can specif-

ically achieve by installing solar panels or by dispensing with a private car, for example. Only over time or if many people act in the same way can the effect also be perceived by the individual.

- Non-sustainable patterns of behaviour can repeatedly be seen in specific spheres of life (at home, at work, in shops, in leisure time, in biosphere reserves). Learning sustainable patterns of behaviour must therefore also take place at many different learning locations.
- Non-sustainable patterns of behaviour are executed by specific players in their various roles and positions (children, adolescents, elderly people, men and women, garden owners, land users, employers, politicians, teachers, etc.). The variety of groups, lifestyles and roles must therefore also be given appropriate consideration.
- Whereas we once assumed that a general environmental awareness would have an impact on all areas of life and all actions, research these days tries to make a distinction between fields of action where non-sustainable behaviour is demonstrated. Because it is far from proven that someone who successfully saves water also buys organic food, dispenses with a car or actively resists the destruction of nature.
- In all of the efforts for "raising awareness" for nature conservation or learning for sustainability, it must always be made clear that terms such as "environment", "nature" as well as "sustainability" and "sustainable development" are constructions of society that it develops, negotiates, questions or confirms in ever new communication processes in society and in scientific and political debates.
- The communication of values and goals of a global, but also locally defined sustainability, is an important prerequisite and component of the process of sustainable development.

But ultimately, it is also a matter of changed patterns of behaviour with a comparably better ecological balance that help to improve living conditions for as many people in the world as possible, while respecting important social and cultural structures. Communicating objectives, increasing knowledge about sustainability and new values are not enough to change patterns of behaviour.

It is not for nothing that there is talk of the gap between knowledge and action. There must, therefore, be different factors that bring about a change in behaviour. For example, they include emotional experiences, such as those encouraged by natural educational science. Social norms in a society or in one's own reference groups, which determine what is "in" and what is "out", what is unacceptable or what is urgently desirable are often also neglected. The empirical man-environment sciences (e.g. environmental psychology and environmental sociology) have now examined a large number of factors that can be considered as obstacles, but also as props for changing non-sustainable patterns of behaviour (Homburg, A., Matthies, E. 1998 and Kruse, L. 2002 b). Corresponding research projects in biosphere reserves have the task of examining these factors and making them of use to further learning processes (Dörner, D. et al.; Stoll, S. 1999). It is now clear that conventional information centres address only some of the aspects relevant to learning sustainability.

Why Learning instead of Education?

Terms such as "education" are much too narrow and too strongly associated with the formal education system and the context of "school". To emphasise that it is a matter of actively changing patterns of behaviour and the values, attitudes, future orientation, motivations, etc. upon which they are based, the term "learning", i.e. "learning for sustainability", should be brought much more to the fore. The term "learning sustainability" also makes it clear that new forms, learning locations and fields of action are required. It is not only schools, kindergartens or specially set up educational establishments that are suitable as learning locations, but also the home, the workplace and the club. In biosphere reserves, learning locations are not only the information centre, but also the market place or the local skilled crafts company.

Sustainable development requires diverse and life-long learning processes.

Sustainable development requires diverse and life-long learning processes that are concerned not only with acquiring abstract knowledge, but also the continuous building and reinforcement of wide-ranging sustainability skills. Learning for sustainability must encourage the acquisition of new, sustainable (e.g. resource-saving, prevention-based) lifestyles (consumption, mobility, living preferences, etc.) for all groups in society in their different living and working situations beyond the narrower educational landscape (school, basic and advanced training). Learning has many facets and the learning offers in biosphere reserves should have a correspondingly diverse shape. Thus, for example, different learning strategies and learning locations need to be selected for the boatman in the Spree Forest than for the commuter from

the Rhön or the forest farmer from Berchtesgaden. New forms of learning, learning media, as well as new partners for the "apprenticeship" (e.g. from the private sector, in the local authorities) need to be found for such extended learning situations.
Learning for sustainability presupposes that in future the employees in the biosphere reserves will see their job as including making the biosphere reserve into a "learning landscape", developing new learning methods and designing learning processes. Learning must be designed holistically and comprise ecological, economic and socio-cultural processes.

Networking and Participation

The UNESCO guidelines make provision for biosphere reserves to aim at close cooperation with education providers (higher education establishments, schools, adult education centres, nature conservation academies, etc.) and existing institutions (associations, museums, professional associations, etc.). But exemplary companies from agriculture and forestry, trade and industry should also be included in the educational programme as further "learning locations".
The Seville Strategy emphasises that the different organisational structures (state and private), educational institutions and local authority Agenda processes are not in competition with each other; on the contrary, they are consciously interlinked and also cooperate with each other constructively (UNESCO 1996). New partners must be acquired and motivated to actively participate in designing the biosphere reserve and to follow the paths towards sustainability. Since learning for sustainability depends on the interaction and communication between the learners, it is thus participatory learning. The World Network of Biosphere Reserves should also be used here to form learning partnerships beyond the local learning landscapes, to encourage exchange processes and communication and, thus, specify the vision for biosphere reserves as "models" for sustainability also for "sustainability learning".

Outlook

With their concept of the various zones, biosphere reserves offer outstanding opportunities for creating learning locations and learning landscapes for learning sustainability and for the introduction of new learning methods for many groups of players every day. Just as the village church, the environmentally friendly B&B hotel and the market place can become learning locations in the biosphere reserve alongside the classic educational establishments (school, higher education establishment, adult education centre), the administrations, skilled craft companies or businesses run on a sustainable basis can be included with new learning offers. The administrative authorities of the biosphere reserves must encourage these partners in actively participating in the process of "learning for sustainability".
Sustainable development must be understood as a process involving comprehensive, worldwide and permanent changes that are repeatedly seen in the specific directly or indirectly environmentally relevant patterns of behaviour of individuals, groups and societies, in lifestyles, patterns of production and consumption in many local contexts. To design these processes of change, many instruments (e.g. financial incentives or levies, laws and administrative rules) must be used, but also education and learning.
Imagine: the biosphere reserve as a sustainability-oriented, exemplary learning landscape – including the possibility of entering into hundreds of connections into the international network of biosphere reserves, to learn from mistakes and the best realised solutions – that really is a model!

Literature

BUNDESMINISTERIUM FÜR UMWELTSCHUTZ, NATURSCHUTZ UND REAKTORSICHERHEIT (1996): Schritte zu einer nachhaltigen Entwicklung: Umweltziele und Handlungsschwerpunkte in Deutschland, Bonn.

DÖRNER, D. (1989) Die Logik des Misslingens. Strategisches Denken in komplexen Situationen.

DÖRNER, D., KRUSE, L. & E. D. LANTERMANN (1995): Umweltbewusstsein, Umwelthandeln, Werte, Wertewandel. Zur Erforschung der Bedingungen und Formen anwendungsorientierten ökologischen Lernens. Begleituntersuchung der Etablierung des Biosphärenreservates Schorfheide-Chorin. In: ERDMANN K.-H., NAUBER J. (1995): Der deutsche Beitrag zum UNESCO-Programm "Der Mensch und die Biosphäre" (MAB), Bonn.

GERMAN MAB NATIONAL COMMITTEE (Ed.) (1996): Criteria for Designation and Evaluation of UNESCO Biosphere Reserves in Germany, Bonn.

HOMBURG, A., MATTHIES, E. (1998): Umweltpsychologie – Umweltkrise, Gesellschaft und Individuum.

KRUSE, L. (1999) Umweltbildung angesichts globaler Umweltveränderungen: Konsequenzen aus umweltpsychologischer Perspektive. NNA-Berichte.

KRUSE, L. (2002a): Lernen für Nachhaltigkeit – nachhaltiges Lernen: Eine ubiquitäre Aufgabe – an vielen Orten, mit vielen Akteuren, in vielen Handlungsbereichen. In: BLK MATERIALIEN ZUR BILDUNGSPLANUNG UND ZUR FORSCHUNGSFÖRDERUNG. Heft 97: Zukunft lernen und gestalten – Bildung für eine nachhaltige Entwicklung.

KRUSE, L. (2002b) Umweltverhalten – Handeln wider besseres Wissen? In: HEMPEL G., SCHULZ-BALDES, M. (Eds.) Nachhaltigkeit und globaler Wandel, pp. 175-192.

STOLL, S. (1999) Akzeptanzprobleme bei der Ausweisung von Großschutzgebieten.

UNESCO (Ed.) (2002): Biosphere Reserves. Special Places for People and Nature, Paris.

3.1.3 Communication and Cooperation

Karl-Heinz Erdmann, Uwe Brendle and Ariane Meier

Introduction

The central task of the administrations of biosphere reserves, together with the people who live and work in these areas, is to draw up and implement exemplary concepts for the protection, maintenance and development of the landscapes represented (GERMAN MAB NATIONAL COMMITTEE 1996). This can only succeed if the population can be motivated to active participation (UNESCO 1996). The key is that all involved are ready and willing to communicate and cooperate.

Those with responsible positions in biosphere reserves should therefore familiarise themselves with the various communicative and cooperative instruments and use them – in the appropriate place. Communication and cooperation are essential elements for the success of biosphere reserves and, thus, a prerequisite for a system of regional development that is compatible with nature.

The population is joining in, as can be seen here with landscape conservation in the Vessertal-Thuringian Forest Biosphere Reserve. The people from the region are often willing to actively support and show commitment to the biosphere reserve.

Communication and Cooperation Services

Communication and cooperation with the various interest groups fulfil political functions (publicity, transparency), social and economic functions (socialisation, orientation, integration, economic gains) and, not least, legitimise the existence of the biosphere reserve.

Communication and cooperation foster the willingness to contribute thoughts and to have a voice as well as to participate in planning and action. Ultimately, communication and cooperation are also necessary for controlling activities in biosphere reserves and for resolving conflicts.

The local and regional interest groups, which also include interested and committed individuals, with their collected knowledge and wealth of experience from the region are also the pool of experts for their region, their home. They are the most important communication and cooperation partners of every biosphere reserve administration.

Various sponsors of social and public institutions already use successful communication and cooperation strategies. The administrations of biosphere reserves in Germany should also develop a strategy with which they can improve the integration of goals and tasks for biosphere reserves in all relevant social, economic and political fields of action in the region.

Communications

The communication concept of D. LUTHE and T. SCHAEFERS (2000) explains which structures are necessary for internal and external communication. It can act as a recommendation for action and a testing instrument in the development of specific communication measures. The concept can also be applied to biosphere reserves:

1) Analysis of the Situation: What are the strengths and weaknesses of the biosphere reserve? On the one hand, the associated external factors (politics, the market, resources, interest groups, goals and responsibilities) and, on the other hand, the organisation of the biosphere reserves (structure, internal relations, communications and image) should be examined.
2) Positioning: Who are we? What do we stand for? It is very important that a guiding principle is formulated here, so as to emphasise the core messages of the biosphere reserves and their concrete plans.
3) Goals: What do we want to achieve? The goals should be worded in a concrete, realistic and flexible manner both within the biosphere reserve administrations and outside.
4) Analysis of the Target Groups: On whom do we want to target? The various support and interest groups, multipliers and collaborators of a biosphere reserve should be defined internally and externally.
5) Content and Messages: What do we want to communicate? In line with points 1 - 4, the messages should be defined and included within the framework of the existing Corporate Identity, i.e. the slogan, image and brand.
6) Strategy: How do we want to proceed? There are three aspects that should be mentioned in particular. Firstly, internal initiatives to promote the acceptance of projects and measures within the administrations of biosphere reserves. Secondly, informative initiatives should be planned for the interest groups and thirdly, intensive and above all personal means of communication should be used with the target groups.

German Biosphere Reserves: Communication and Cooperation Structures

The communication and cooperation activities of the biosphere reserves in Germany aim at making UNESCO biosphere reserves known to the broad public. A positive image is to be developed. The net of areas should be promoted to attract more attention as model regions for sustainable development – nationally and internationally.

- Working Group of the Biosphere Reserves in Germany (AGBR/EABR)

Initiated by the Federal Ministry for the Environment, Nature Conservation and Nuclear Safety (BMU), representatives of the German biosphere reserves and external experts have met regularly to intensely exchange views, information and experiences since 1990. The group develops strategies and standards and edits publications.

- MAB/MAB-National Committee Germany

The MAB Programme provides the framework, sets international standards and ensures international transfer of experiences and knowledge. The German MAB National Committee is contributing to the further development of the Statutory Framework of the World Network of UNESCO Biosphere Reserves as well as of the national Criteria for Designation and Evaluation of UNESCO Biosphere Reserves in Germany and documents periodically the development of German biosphere reserves in line with UNESCO requirements.

- EUROPARC Deutschland/EUROPARC Federation

EUROPARC as the umbrella association of all large-scale protected areas (i.e. biosphere reserves, national parks and nature parks) serves the communication among the administrative bodies involved. By public relations work and lobbying as well as by transfer of know-how EUROPARC is promoting biosphere reserves nationally and internationally.

- Image

The members of EUORPARC Deutschland have created a common image for all large-scale protected areas. This is supported by the biosphere reserves in Germany and actively implemented. The common image serves the recognition, increases acceptance and further forms the common appearance.

- Guiding principles

On the basis of the General Guidelines for Biosphere Reserves in Germany (AGBR, 1999, cf Chapter 2.4) the umbrella association EUROPARC Deutschland has drawn up guidelines for biosphere reserves, national parks and nature parks respectively. They serve to sharpen the profile of the different categories externally and to improve the internal cooperation.

7) Measures: Which concrete means of communication should we implement? A range and combination of different measures and a detailed schedule is required here.
8) Responsibilities, Tasks and Resources: Who should take on responsibility? This should be made very clear. It is wise to separate project management, organisation and implementation and to coordinate accurately. During implementation it is essential to have a constant exchange of information (such as regular meetings, e-mail circulars, etc.) and checklists for planning measures and schedules, including responsibilities.

Cooperation

Communication facilitates successful cooperation by providing official benefits such as transparency, authorisation, neutrality and unofficial requirements such as credibility, honesty and trust (Vieth, C. 2000: 159). Different methods and models exist to promote active cooperation (including round tables, presentation procedures, planning areas, workshops, conferences and forums). When choosing a method, consideration should be given to the concerns, the general requirements and the participating organisations.

As with communication, there are initiatives to generate cooperation systematic and professional planning, implementation and follow-up evaluation. The political "components" formulated by U. Brendle (1999) provide important support in this regard. Goals and concerns can only find support among committed people, leading members and influential advocates of biosphere reserve interest groups. Continuous and competent communication and personal contact is essential.

Successful biosphere reserve managers have the ability to communicate, managerial and strategic capabilities, social and political knowledge and tactical skills at their disposal. The ideal conditions for success include sufficient resources, such as personnel and money. Comprehensible projects (simple structures, a limited number of participants) promote successful cooperation.

The management of a biosphere reserve can acquire partners for successful cooperation if it reacts to the pressure stemming from increasing ecological, social, political and economic problems and offers appropriate solutions through cooperation. "Winners' coalitions" made up of players with different interests and demands can be very important. They will succeed if all involved reap the benefits of cooperation. Early successes are important because they prove that the biosphere reserve administrations, as well as the groups – such as sponsoring bodies and advisory councils – that cooperate very closely with them, are efficient and able to act very effectively. They also foster acceptance within the interest groups.

Improvement in the cooperation between the individual biosphere reserves is also important for their success. Particularly in the areas of public relations work and nature education, the biosphere reserves must agree upon their goals, methods and concepts, share their collective experiences and present the "biosphere reserve" concept consistently. For example, within cooperation the biosphere reserves can share results from various fields and staff can be seconded (for a limited time) or they can implement joint campaigns and measures. This is especially true for the international sphere.

As a rule, projects in the area of communication and cooperation run dynamically. Even during implementation, changes should be reacted to with flexibility. It is essential to learn from mistakes. This promotes project success and also stabilises the cooperation between the various interest groups and the administrations of biosphere reserves.

Literature

BRENDLE, U. (1999): Musterlösungen im Naturschutz – Politische Bausteine für erfolgreiches Handeln, Bonn.

GERMAN MAB NATIONAL COMMITTEE (Ed.) (1996): National Criteria for Designation and Evaluation of UNESCO Biosphere Reserves in Germany, Bonn.

LUTHE, D., SCHAEFERS, T. (2000): Kommunikationsmanagement – Strategische Überlegungen und konkrete Maßnahmen für eine beziehungsorientierte Öffentlichkeitsarbeit. In: NÄHRLICH, S., ZIMMER, A. (Eds.): Management in Nonprofit-Organisationen. Eine praxisorientierte Einführung. Opladen, pp. 201-223.

VIETH, C. (2000): Wege zur besseren Akzeptanz. In: ERDMANN, K.-H., KÜCHLER-KRISCHUN, J. & C. SCHELL (Eds.): Darstellung des Naturschutzes in der Öffentlichkeitsarbeit. Erfahrungen, Analysen, Empfehlungen. BfN-Skripten 20, pp. 157-163.

3.2 Conservation of Nature and Landscape

3.2.1 Objectives and Strategies for Nature Conservation

Michael Vogel

UNESCO was one of the first international organisations to recognise the worldwide challenges resulting from human's increasingly intensive interventions in the balance of nature. As a result, nature conservation became an issue of survival for humanity. The importance of nature conservation and protection of the environment therefore motivated UNESCO to set up the international Man and the Biosphere Programme (MAB) in 1970. One of the main goals was to examine the area of conflict between humans and the environment and to demonstrate ways to bring about a lasting improvement in this relationship. Nature conservation should not only be geared towards the needs of humans. Nature should also be protected for its own sake, so that opportunities remain for future generations. This ethical concern for the future will also receive lasting support from the MAB Programme.

Unspoilt nature: impressions of Vilm Island, the core area of South-East-Rügen Biosphere Reserve

Nature conservation is a very complex field of work with many inter-related individual aspects which need to be taken into consideration. In general, goods worth being protected, such as endangered species, communities, processes and activities within nature and also beauty, diversity and the pecularity of nature and landscapes should be safeguarded. Conservation and preservation, sustainable development and research,

education and environmental monitoring form the three pillars on which biosphere reserves should grow into model regions of sustainable development. Embedded within the frameworks of a large range of commitments (from supranational to national), conservation work should change its focus from dealing with individual problems associated with a particular area to an approach that takes the total area and overall causes into account.
The content and scope include:

Development and Application of a Cooperative and Constructive Process for the Concepts Designed for the Use and Management of Land

The concepts designed for the use and management of land should protect and be compatible with the processes and activities within nature.
From the outset, the objectives of nature conservation should be implemented on 100 percent of the biosphere reserve area. The total area and overall causes consequently need to be taken into consideration. Nature conservation work should be an integral part of using the land. This means that those active in nature conservation work must intervene in the distribution and development issues of society. To this end, the players in conservation must formulate models, quality objectives and possible guidelines. In addition, the players could introduce mechanisms to control the balance of interests between social groups. The principle of sustainable use, perhaps better expressed as the factor that determines the general conditions for use, must be regarded as the main priority. The UN Conference on Environment and Development in Rio de Janeiro in June 1992 was indeed supposed to represent such a starting point. "Agenda 21", agreed at the Conference, is an extensive, dynamic action programme that contains detailed instructions for environmental and developmental political action. To this end, the public worldwide needs practical examples that put the ideas from the Rio Conference into practice. These examples only work, however, if they consider the social, cultural, intellectual and economic needs of society and if, at the same time, they are supported by a solid scientific basis. In 1995, the International Conference on Biosphere Reserves in Seville, Spain, therefore confirmed that biosphere reserves represent such examples, particularly because they demonstrate ways to achieve a more sustainable future.

There are various aspects of sustainability and sustainable use for nature conservation work:

- sustainable use from an ecological point of view
- sustainable use from an economic point of view
- sustainable development from a social point of view
- sustainable use from a spatial point of view
- sustainable use from the point of view of time

Systems for Protected Areas

It is a well-known fact that all species can be preserved in landscapes if approximately 25 per cent of the area consists of natural and semi-natural areas. The objective must therefore be to set up different "area protection categories" that have an effect on the total area. The core areas, be it an individual area or of mosaic structure, must belong to the strongest protection category. It is worth striving for a future biotope network comprising existing conservation areas, linked to areas managed contractual nature conservation and areas that are subject to other legal regulations (e.g. water legislation). Within this structure, nature should hold greater priority than culture. In other words, the consecution of different natural conditions in the same area must be ensured (i.e. succession) to enable individuals and species to live alongside one other (i.e. diversity of systems).

Animals need conservation: Black Woodpeckers (Dryocopus martius) in Vessertal-Thuringian Forest Biosphere Reserve

The goal should be to protect the basic functions of the ecosystem that preserve and support natural dynamic processes, such as area and population changes, new settlements, succession, speciation and evolution under undisturbed conditions. Moreover, this should contribute to a careful controlling of the use of the landscape, in the sense of promoting techniques that are sustainable, nature-friendly and save energy and materials (wise use, sustainable use).

The Value of Nature – Steps towards Its Economic Valuation

"Nature" includes all of the world's living systems: ecosystems, plants and animals, landscapes, biological diversity. No human culture or society could survive without the multitude of services these systems provide. Intact ecosystems purify air and water and regulate plant and animal populations. Natural products are the basis of our food production as well as of considerable parts of industrial production. Genetic resources are indispensable in agriculture, forestry, fisheries, medicine and pharmacy. Without natural pollinators, agriculture would be impossible. Natural landscapes are destinations for recreation and tourism. In an article in the scientific journal Nature in 1997 R. Costanza et al. calculated the total value of the ecosystem services of natural and near-natural ecosystems of the whole planet as between 16 and 54 million million US$ annually, about twice the size of the gross global product!

Considering that nature has such an immense value for our own survival, one would expect humans to treat her with utmost respect. Often enough, this is not the case at all. How can this discrepancy be explained? One of the reasons is that in many cases there is no direct economic value that can be attributed to nature, although it is obvious that nature is not worthless. The value of nature finds only inadequate expression in economic calculations and societal decisions. Often nature has the status of a public good that does not cost anything and is freely accessible. In many cases, any use that is made of this good brings about an immediate advantage, but entails a disadvantage in the long run, the effects of which will have to be borne by the next generation. Or the individuals or groups who stand to gain from some use of nature are not those that will have to bear the associated cost. Therefore, making use of nature often produces "hidden costs".

The question of how the services and products of nature can be made to acquire a calculable "price" is dealt with in a fairly new field of research called "environmental economics". Environmental economics makes use of a variety of valuation methods to calculate or at least estimate the hidden value of nature. A cost-benefit analysis, for instance, is used to calculate the total economic cost by subtracting from the total value of the benefits gained by a specific use of nature the value of all the costs that this use entails. Benefits and costs can be quantified either by market prices or have to be estimated by other means, e.g. a "willingness to pay" analysis (through interviews among target groups a monetary value is found that this group is willing to pay for a specific service of nature or for excluding a certain use); a value may be given for the cost of substituting a service of nature by purely technical means; or values gained in comparable studies are taken as a first approximation. The cost-benefit analysis of a specific use of a given ecosystem must also take into account the indirect effects of this use. Take as an example a river that will be changed by infrastructure development. The losses caused among the fisheries in this river will have to be added to the costs for infrastructure development in order to know the total cost produced by this measure.

The economic valuation of nature is certainly no panacea against a non-sustainable utilisation of nature. However, these methods can be of great help in demonstrating the value of nature among the public and among decision-makers. They can help us to decide between different uses of nature, to identify priorities, to assess the implications of infrastructure development and to choose between instruments of nature conservation. Their greatest disadvantages lie:

- *in the difficulties of integrating differing time frames (short-term benefits outweigh long-term and even severe disadvantages),*
- *in the uncertainties concerning future and as yet undiscovered potentials of utilising nature,*
- *in the necessity of applying the precautionary principle even when possible additional costs cannot be quantified,*
- *in the impossibility of substituting many ecosystem services and products by technical means and*
- *in the difficulty of quantifying ethical, moral, cultural, scientific, religious and other values which can be indefinitely high for certain groups.*

Methods for the economic valuation of nature are already being applied in a large number of fields ranging from the calculation of equalisation levies intended to counteract impacts on nature in Germany to carbon offset mechanisms and emission trading that are to be implemented after the Kyoto Protocol has entered into force. However, this approach still has a much larger potential for demonstrating the value of nature to all of us.

Rudolf Specht, Federal Agency for Nature Conservation (BfN)

Conservation of Individual Areas

The conservation of individual areas should adhere to special criteria, such as:

- conservation of endemic species that are only found in Central Europe;
- conservation of natural landscapes with long periods of regeneration and/or development, such as high mountain regions, river landscapes and mud flats;
- cultural landscapes with an anthropologically controlled diversity. The history of civilisation should be the main consideration here;
- conservation of biotope types, which, if at all, can only be reproduced with great difficulty, such as areas poor in nutrients and old/mature and/or wet, damp/dry areas.

An area of land which offers space for minimum viable populations of species and which is sufficient in size to reactivate independent processes must be guaranteed.

Nature conservation areas, as here in the South-East-Rügen Biosphere Reserve, are clearly marked by signs.

Research, monitoring and efficiency control

Nature conservation is a discipline based on social values and can only carry out its legislated responsibilities on a scientific basis. In addition, agreed, targeted research, based on complex objectives is necessary for defining, assessing and solving problems. Complexes of objectives for biosphere reserves include, for example:

- research into changes in ecosystems, brought about by human activities and the consequences of these changes for people;
- identification and comparison of the structure, function and dynamics of natural, modified and cultivated ecosystems;
- research and comparison of dynamic interactive relationships between ecosystems and socio-economic processes;
- definition of scientific criteria for the sustainable management of natural resources.

The MAB Programme deals with even more than this. It emphasises the necessity of integrated, interdisciplinary, not only multidisciplinary research. Cooperation is needed between nature, social, economic, arts, planning, engineering, agricultural and forestry disciplines. Furthermore, the national income derived from different areas will need to be included. The goal here is to take action that will lead to social monitoring. In addition, when there are budgetary constraints, it is essential to repeatedly check and demonstrate the efficiency and effectiveness of the control measures in place.

Environmental Education, Public Relations and Communication

All knowledge and research is and will remain useless if it is limited to experts. Effectiveness and the ability to act is only achieved if it is exposed to people through public relations work, environmental education and communication. A new relationship must also be established between humans and nature. It must be made clear that nature and the environment directly concern the most limited, personal aspect of every individual's life and that subjective processes, harmful to nature, which are not always noticeable and on-going, affect the entire existence of every individual in his or her interaction with nature. The mechanisms and methods of dealing with this should be applied and tailored to all age groups, from children in the pre-school age group to adults in the widest range of professional groups.

Opportunities for the Future

There should be a change of direction within nature conservation work and nature conservation policy. Nature conservation work, as an action-focused, applied, valued discipline focusing on specific areas of land, must achieve the same "status" as other disciplines that work with land and nature. The aim should therefore be to develop the whole process so that it acts as a mediator and moderator, taking on the responsibility of defining frameworks and regulations. The future focus of sustainable conservation work must be to initiate and present such processes of negotiation and to provide them with the input required to scientifically evaluate the situation and assess the consequences of decisions.

Literature on the Subject

BMU (Bundesministerium für Umwelt, Naturschutz und Reaktorsicherheit) (Ed.) (1992): Bericht der Bundesregierung über die Konferenz der Vereinten Nationen für Umwelt und Entwicklung im Juni 1992 in Rio de Janeiro.

Costanza, R., d'Arge, R., de Groot, R., Farberk, S., Grasso, M., Hannon, B., Limburg, K., Naeem, S., O'Neill, R. V., Paruelo, J., Raskin, R. G., Sutton, P. & M. van den Belt (1997): The Value of the World's Ecosystem Services and Natural Capital. In: Nature 387 (15 May 1997), pp. 253-260.

Deutscher Bundestag (Ed.) (1998): Abschlussbericht der Enquete Kommission "Schutz des Menschen und der Umwelt – Ziele und Rahmenbedingungen einer nachhaltig zukunftsverträglichen Entwicklung" des 13. Deutschen Bundestages: Konzept Nachhaltigkeit. Vom Leitbild zur Umsetzung.

Deutsches MAB-Nationalkomitee (Ed.) (1990): Der Mensch und die Biosphäre; Internationale Zusammenarbeit in der Umweltforschung.

Deutscher Rat für Landespflege (2002): Gebietsschutz in Deutschland: Erreichtes – Effektivität – Fortentwicklung. In: Schriftenreihe des Deutschen Rates für Landespflege, Heft 73.

Erdmann, K.-H., Spandau, L. (1997): Naturschutz in Deutschland: Strategien, Lösungen, Perspektiven.

Hammond, A., Adriaanse, A., Bryant, D. & R. Woodward (1995): Environmental Indicators: A Systematic Approach to Measuring and Reporting on Environmental Policy Performance in the Concept of Sustainable Development. World Resources Institute.

Kastenholz, H. G., Erdmann, K.-H. & M. Wolff (Eds.) (1996): Nachhaltige Entwicklung; Zukunftschancen für Mensch und Umwelt.

Lass, W., Reusswig, F. (Eds.) (2002): Social Monitoring: Meaning and Methods for an Integrated Management in Biosphere Reserves. Report on an International Workshop. Rome, 2 - 3 September 2001. Biosphere Reserve Integrated Monitoring (BRIM) Series No. 1.

Nationalparkkommission der IUCN (CNPPA) (1994): Parke für das Leben: Aktionsplan für Schutzgebiete in Europa.

Plachter, H., Bernotat, D., Müssner, R. & U. Riecken (2002): Entwicklung und Festlegung von Methodenstandards im Naturschutz. In: Schriftenreihe für Landschaftspflege und Naturschutz, Heft 70, Bundesamt für Naturschutz.

Primack, R. B.: (1995): Naturschutzbiologie.

Remmert, H. (1992): Ökologie: ein Lehrbuch.

Scherzinger, W. (1996): Naturschutz im Wald: Qualitätsziele einer dynamischen Waldentwicklung.

Schweppe-Kraft, B. (2000): Innovativer Naturschutz – Partizipative und marktwirtschaftlichen Instrumente. In: Angewandte Landschaftsökologie, Heft 34, Bundesamt für Naturschutz.

UNESCO (Ed.) (2002): Biosphere Reserves: Special Places for People and Nature.

Zeitschrift für Angewandte Umweltforschung (1994): Umweltdiskussion: Sustainable Development, ZAU Jg. 7, Heft 1.

3.2.2 Cultural and Natural Landscapes and the New Wilderness

Michael Succow

Starting Situation

Natural landscapes, in Central Europe primarily deciduous forests, initially evolved into semi-cultural formations and then into historically developed cultural landscapes with their diverse structures and beautiful landscapes. In the second half of the last century they were transformed into today's predominant "production landscapes". Natural landscapes in Germany can only be found scattered in the high mountains. The last semi-cultural formations (heaths, wetlands, grazed forests, oligotrophic grassland communities) are preserved by means of managed use in nature conservation areas.

Historically developed "harmonious" cultural landscape - an environmentally sensitive area: safeguarded in the "Mecklenburg Switzerland" Nature Park

The current land use that shapes the vast majority of our agricultural landscape is characterised by enormous substance imports and a high expenditure of outside energy. The consequences of this are eutrophication, the accumulation of pollutants, fragmentation of the landscape and a total mechanisation with the result of drastic cuts in jobs. The intensive usage forms solely geared towards the production of food and raw materials lead to a dramatic loss in biodiversity and the functioning ability of the natural balance. The term "cultural landscape" hardly applies to these "homogenised" production landscapes solely geared towards maximising yields, with their humus-depleted and compacted soils, weed-free monocultures and extremely reduced crop rotation any longer (Haber, W. 1998, Succow, M. et al. 2001)!

"Modern" production landscape with drum irrigation system: ground moraine plain in the Uckermark region. The heterogeneity of the soil due to heavy soil erosion that has occurred after just a few decades of intensive use is clearly visible.

Former semi-cultural landscape formations like shrub pastures can only be preserved in nature conservation areas by maintenance measures, e. g. grazing by Swedish Fjaell cattle in Müritz National Park.

At the same time, the need to live "in the countryside" and to recuperate in "pristine nature" is growing in an industrial, service and leisure society increasingly marked by overproduction and urbanisation; in other words, the importance of the countryside as a human habitat is growing.

Cultural Landscapes of the Future

The conservation of rural areas as cultural landscapes will be achieved in the future only by means of a more ecologically and socially oriented land use policy. Future guiding principles will have to include placing a value on ecological benefits or the creation of markets for nature conservation services. Safeguarding the natural balance, i.e. the ability of the ecosystems to function independently, must be given priority in this connection. The preservation of larger sections of the landscape as historical cultural landscapes is another key point.

Further price collapses for agricultural products and the expected general reduction in transfer payments mean that the current concepts, i.e. every advancing intensification, are coming to their limits.

But what is to happen in future with the unfavourable sites, the so-called problem areas in the countryside, i.e. agricultural locations in medium-range mountains, the sandy soils and lowland sites that can only remain usable for conventional agriculture with constant hydromelioration and the costly maintenance of dikes and pumping stations? After all, these so-called marginal sites currently account for up to a third of the total agricultural land in some *Länder* (Bork, H. R. et al. 1998). This proportion will increase.

One sensible strategy is certainly the designation of natural development areas, in other words, landscapes that remain unused and are returned to nature's own dynamics where wilderness will return. Here, man will only be an observer, admirer and learner. In landscapes with not much woodland, for example, large areas should be surrendered to reforestation. But more urgently than ever we need model landscapes that fulfil all of the functions of a healthy cultural landscape and that serve the aims of environmentally friendly and socially compatible regional development. They form an ecologically and socially important counterweight to urban settlements and, naturally, also to the current production landscapes. The conservation and the development of extensive and/or alternative cultural landscapes is one of the main tasks of biosphere reserves. In general, they are landscapes that have a nationally significant wealth of nature due to their large size, low levels of fragmentation and characterisation by cultural landscapes and they should be defined as national natural and cultural heritage. Moreover, in the buffer zones and transition areas of biosphere reserves, humans are is at the heart of considerations and, thus, economic and social development go hand in hand with the conservation of the most valuable landscapes.

Changing the General Conditions

Obviously, the question of the long-term affordability of more extensive use of the landscape is also raised. It is certainly not economic under the approaches of the current general economic conditions. But this is just as much the case for the favourable sites with their especially high transfer payments. However, if we commit ourselves to sustainable development – in other words to the interlinking of ecological, social and economic needs as the German Advisory Council of Experts on the Environment laid out especially clearly in its report of 1994 and as the MAB Programme specified over 30 years ago – it is necessary to change the "general conditions". In the past there were hardly any incentives for land users to perform ecological services. The one-sided orientation of the price system to agricultural products led to an increase in the so-called negative external effects of land use. However, the classic agricultural and forestry economy does not offer any approaches to solve these problems (HAMPICKE, U. 2000 a and b). An "eco-tax" system with levies for negative ecological impacts – e.g. a nitrate tax, a pesticide tax, an import feed tax, as well as the removal of the subsidies for maize cultivation and diesel privileges – would bring us closer to the model of land use with low environmental impact based on life cycles. Prices reflecting the "ecological truth" would have to be achieved for "organically" grown products. They should be used as a basis for a reference system to pay for ecological services. Today, more than ever, it is a matter of closed substance and energy cycles at farm level. By contrast, the previous price systems, aiming only at increasing production efficiency, even offer incentives to damage natural commodities, associated with subsequent limited payments to reduce this damage.

Land use in the 21st century must be more socially compatible and environmentally friendly. In particular, healthy, permanently usable landscapes that are ecologically and socially intact, that stabilise the living space in which the carrying capacity of neither nature nor the rural community is exceeded, are needed around the large conurbations due to the urbanisation that is increasing all over the world. Planning policy, land use policy, water management policy, nature conservation policy and socio-economic development in rural areas are irrevocably linked to each other. Ultimately, they must be understood as a unit.

In the meantime, we have to implement organic farming, natural forestry, forest-friendly game management, natural water use and even environmentally friendly tourism on 100 per cent of the land and not just in biosphere reserves. This would make it possible to let more people take sensible action. In this connection, we could produce sufficient quantities of high-quality food and would not need any expensive storage for the overproduction of foodstuffs. The gigantic energy use would be stemmed, goods streams reduced and the long overdue "world compatibility" introduced. Townspeople, too, would regain a connection to their local landscape. Even now, a broad majority of the population would probably agree to transfer payments for ecological services in conjunction with high-quality food and good groundwater. If the real ecological price were to be paid for the subsidised traffic and the too cheap mobility, we would not need to worry much at all about many rural areas. Production, processing and marketing in the region would then be normality again. Goods transported from far away would become luxury items. Local trade would blossom again. Production and consumption would belong together again. More people would have jobs again. Setaside would no longer be on the agenda!

Healthy cultural landscapes keep the equilibrium between change and preservation. Their conservation and/or their recreation require the interaction of many partners, such as farmers, foresters and nature conservationists, entrepreneurs and representatives of the transport and tourism industry, architects and landmark conservationists, as well as representatives of churches and cultural life.

Intact cultural landscapes can be viewed as alternative models for the urbanised world; after all they are areas in which humans have developed their culture in such a way that nature has been able to develop a great wealth in spite of, and sometimes because of, the use. In these areas, people find mental and spiritual well-being, artistic inspiration, creativity and hope in an age increasingly characterised by a rootlessness, a lack of ties and orientation. But they are also returning to religiosity, awe of nature, more modesty.

Cultural landscapes are an expression of an interaction between humans and nature, of cultural and biological evolution. Cultural landscapes form the key to an ecologically and culturally adapted use of nature. They are of outstanding importance for the implementation of the concept of sustainable development, the only viable path for human civilisation in the future. (DÖMPKE, S., SUCCOW, M. 1998, SUCCOW, M. et al. 2001).

Natural Development Areas – Daring to Return to the Wilderness

To really experience wilderness in Europe we have to go to the extreme peripheral areas. Wilderness in the strict sense, including "wild animals", still exists only in the far north of the continent, on Spitsbergen, in Lapland, Northern Russia, and in the peak areas of high European mountains between the Pyrenees, Scandinavia and the Caucasus.

But in Central Europe, too, large areas of the landscape have to be left to nature or be returned to it. In 1994 the report of the Federal Government's Advisory Council of Experts on the Environment called for at least five per cent of the land area in Germany to be reserved – this was quite revolutionary at the time, but a figure of ten per cent is often discussed nowadays.

Since the early 1990s, "wilderness" has also been under discussion as a guiding principle in nature conservation. Merely the fact that "daring to return to the wilderness" appears to be necessary shows how deep-seated the fear of wilderness is in the public consciousness. At the beginning of the 21st century, "wilderness" will become the major challenge for nature conservation.

The removal of the marginal sites in Central Europe from land use opens up the opportunity to leave at least five to ten per cent of the land to nature's own dynamics. Use here can be maintained only by society paying especially high transfer payments. By contrast, natural development areas cost hardly any money, just an undertaking by society to consciously renounce material use, an pledge to "untamed" nature – ultimately for the benefit of us all. National parks and core areas as wilderness areas in biosphere reserves are the relevant categories of protection for this. They can provide astonishingly economic yields for the population living in the transition areas. The social component is also significant: although areas in which wild nature can be experienced do not need any foresters or farmers, fish managers or water managers, they do need nature conservation managers and nature interpreters, in other words landscape managers. This is the start of a new profession, in which a great deal of the knowledge and skills of the old professions of farmer, forester, etc. are upheld.

Obviously, this raises the question as to which landscapes in our densely populated Central Europe will still be available for such a scheme in the future.

Post-mining landscape in Nochten/Lausitz that has not yet been recultivated, with a "residual pit" after completion of coal removal

Medium-range mountains as well as low-nutrient, dry sandy landscapes, stony terminal moraine landscapes that have strong reliefs and very changeable soil quality and, increasingly, plain sites that once had a regulated water balance (moors, coastal plains and alluvial plains) are particularly affected by this removal from use.

The large military exercise sites that are no longer used can also be redefined as natural development areas. They open up the opportunity of leaving large areas to nature's own dynamics. In the new *Länder* alone, 3.5 per cent of the land area is taken up by military exercise sites. They are a once-in-a-lifetime potential of landscape that is becoming free (Deutscher Rat für Landespflege 1993, Gorissen, I. 1998, Beutler, H. 2000).

Finally, we should also have the courage to surrender to nature the post-mining landscapes in eastern Germany that have not yet been recultivated. In many respects, these extensive mining areas are reminiscent of landscapes in the early post Ice Age era. The bare sand, gravel and clay areas will gradually mature in the spaces full of life by means of "natural succession".

Even now, lakes and moors, sandy heaths and the first primeval forests have sprung up here. In 100 years our grandchildren and great grandchildren will be able to experience new wilderness areas that have developed without any human intervention. In the Upper Lausitz Heath and Pond Landscape Biosphere Reserve, a former, non-recultivated lignite mining area has been included in the core area.

The new wilderness areas with a great emotional impact will certainly attract tourists and artists as well as scientists. This means that our urbanised society will develop new values, which – without a doubt – can be natural islands of wilderness of increasing importance, particularly in the environs of the major conurbations of Halle/Leipzig and Berlin/Cottbus. So we must seize the last opportunity and safeguard for nature conservation as much of the post mining landscapes as possible that have not yet been recultivated, as the core areas of biosphere reserves and/or national parks (Succow, M. et. al. 2001).

It has been and still is a long path of understanding to see that human civilisation can only continue if we preserve the ability of the foundations of our lives to function – the natural space available to us – and change them as little as possible. This means finding and implementing sustainable forms of land use and leaving or returning to nature's dynamics natural areas that are unused or no longer needed for their own sake. In this vein we quote the writer Reimar Gilsenbach (1925-2001): "If we leave nature unchanged, we cannot exist, if we destroy it, we perish. In the long run, the perilous, narrowing tightrope walk between changing and destruction will be achieved only by a society that accepts ecological principles and whose ethics are to be at one with nature".

Literature

Bork, H. R., Bork, H., Dalchow, C., Faust, B., Pior, H.-P. & T. Schatz (1998): Landschaftsentwicklung in Mitteleuropa: Wirkungen des Menschen auf Landschaften, Gotha/Stuttgart: Kett-Perthes.

Beutler, H. (2000): Landschaft in neuer Bestimmung – Russische Truppenübungsplätze, Neuenhagen: Findling-Verlag.

Deutscher Rat für Landespflege (Ed.) (1993): Truppen-

übungsplätze und Naturschutz. In: Schriftenreihe des Deutschen Rates für Landespflege 62.
DÖMPKE, S., SUCCOW, M. (1998): Cultural Landscapes and Nature Conservation in Northern Eurasia. – Proceedings of the Wörlitz Symposium, March 20-23, 1998. Edited by Naturschutzbund Deutschland (NABU) in cooperation with The Nature Conservation Bureau and AID Environment, Bonn.
GORISSEN, I. (1998): Die großen Hochmoore und Heidelandschaften in Mitteleuropa, Siegburg: Selbstverlag I. Gorissen.
HABER, W: (1998): Von der Kulturlandschaft zur Landschaftskultur. Festvortrag anlässlich der Festveranstaltung "Verleihung des Alternativen Nobelpreises" an Prof. Dr. M. Succow. In: Greifswalder Universitätsreden, Neue Folge Nr. 85, pp. 26-41.
HAMPICKE, U. (2000 a): Naturschutz – ökonomisch gesehen. In: ERDMANN, K.-H., MAGER, T. J. (Eds.): Innovative Ansätze zum Schutz der Natur, Visionen für die Zukunft, Berlin, etc.: Springer Verlag, pp. 127-150.
HAMPICKE, U. (2000 b): Möglichkeiten und Grenzen der Bewertung und Honorierung ökologischer Leistungen in der Landwirtschaft. In: Schriftenreihe des Deutschen Rates für Landespflege 71, pp. 43-49.
RAT DER SACHVERSTÄNDIGEN FÜR UMWELTFRAGEN DER BUNDESREGIERUNG (1994): Umweltgutachten 1994 – Für eine dauerhaft-umweltgerechte Entwicklung, Stuttgart: Metzler-Poeschel Verlag.
SUCCOW, M., JESCHKE, L., KNAPP, H. D. (2001): Die Krise als Chance – Naturschutz in neuer Dimension, Neuenhagen: Findling-Verlag.

3.2.3 *Cultural Landscapes and Biodiversity*

Harald Plachter and Guido Puhlmann

Europe's Contribution to a World Nature Conservation Strategy

Europe lost parts of its "untouched" nature thousands of years ago. All European landscapes have been ecologically shaped by humans, if not structurally, then through other influences, such as hunting, water management or inputs of substances from the atmosphere.
We must assume that with a growing world population there will soon be many more areas of this kind. "Nature conservation in used landscapes" is thus the strategy that will very quickly gain global significance in our modern societies.
Possible solutions will always be the result of a "trial and error" principle due to the immense complexity of nature. On this basis, no continent has more experience in balancing human use interests against ecological restrictions than Europe. For thousands of years, European cultures have been using this principle of "trial and error" to find forms of use that satisfy their needs for life and, at the same time, do not overload the ecological regulating mechanisms. In the process, many mistakes have been made for which European societies are still paying. Many areas have been overused, e.g. the karst landscapes in the Mediterranean area or the heaths of Central Europe and southern England. But no continent has collated more information on the "limits of growth" over millennia. European know-how could thus be the key to precisely defining sustainability in other regions of the world, especially as land use techniques of European origin are largely used there now.

Landscape

"Landscape" is a term that everyone understands – or at least believes to understand. But there is hardly a term that is used in the sciences and in the public with so many different meanings as this one: a garden designer will understand landscape to mean something completely different from an ecologist, an urban dweller something completely different from an inhabitant of the sub-arctic tundra. For a painter a landscape only becomes reality when he has "realised" it, including his cultural values (cf landscape painting) on the canvas, for the natural scientist it is the web of ecological functions at a high spatial level. A white-tailed eagle would certainly define "its" landscape very differently from a beetle.

Regardless of all scientific exactness: landscape is obviously a term that links, a term that facilitates communication between people of very different origins, interests and values. Even if landscape exists only as a manifestation of human values: a distinction between "natural landscapes" (natural ecological processes predominate and determine the space-time structure) and "cultural landscapes" (human influence dominate both in processes and in structures) obviously makes sense.

Biological Diversity

Cultural landscapes are far removed from a "natural" landscape and make only comparatively small contributions to the basic principle of "conserving the naturalness" in nature conservation. If designed accordingly, however, they can significantly contribute to another basic principle, namely to "preserve a high level of biodiversity". And – subject to adequate substance and energy inputs by humans – they can be just as "stable" as natural landscapes in the same place.
However, in terms of their ecological properties, they are clearly distinct from many natural landscapes.
In natural landscapes, species diversity is often already very high in small areas (tropical rain forests, coral reefs). Several thousand species per hectare are not a rarity. In cultural landscapes, by contrast, many habitats are comparatively poor in species. The difference lies in the fact that many natural landscapes are made up of basic units that are always the same or similar, whereas cultural landscapes often offer a mosaic of very different habitats. In small areas, natural landscapes are consequently often much more biodiverse, in large areas the differences are blurred.
The question as to what is "more valuable" in terms of "diversity" is thus primarily a spatial problem of scale. With respect to the individual ecosystem, the answer is undoubtedly the natural landscapes, with respect to the landscape it is not infrequently the cultural landscapes (with the exception of large vertebrates).

Cultural Landscapes

Since 1992 the category of "cultural landscape" has been added to the UNESCO World Heritage Convention. A pragmatic definition had to be found. The result may appear worthy of scientific discussion. But for practice, it has proved to be sensible for over a decade now (cf box).

Categories of cultural landscapes according to the "Operational Guidelines" of the UNESCO World Heritage Convention

Operational Guidelines, Paragraph 39:
Three main categories:

1. (...) clearly defined landscape designed and created intentionally by man. This embraces garden and parkland landscapes constructed for aesthetic reasons which are often (but not always) associated with religious or other monumental buildings and ensembles.

2. (...) the organically evolved landscape. This results from an initial social, economic, administrative and/or religious imperative and has developed its present form by association with and in response to its natural environment.
Two sub-categories:
- *(1) Relict (or fossil) landscape: evolutionary process came to an end but distinguishing features are still visible*
- *(2) Continuing landscape: retains an active role in contemporary society*

3. (...) the associative cultural landscape. (...) Is justifiable by virtue of the powerful religious, artistic or cultural associations of the natural element rather than material cultural evidence

(...)

Case Study: Dessau-Wörlitz Garden Kingdom in the Middle Elbe River Landscape Biosphere Reserve (Type: "Designed landscape")

The cultural landscape of Dessau-Wörlitz Garden Kingdom lies in the area around Dessau in the Federal *Land* of Saxony-Anhalt. It originated in the late 18th century as a complex reform work by Prince Leopold III Friedrich Franz of Anhalt-

Dessau. With the aim of finding a way out of poverty and crisis, under the conditions of the time, the ruler encouraged the use of the landscape, improved the education and social conditions of the population and designed the landscape according to aesthetic aspects. The guiding principle was to combine "the beautiful with the useful" in line with Rousseau's thoughts.

Today, we would say that the principles of sustainable development were applied here 200 years ago in a region or even a state. The Garden Kingdom is thus suitable – more than hardly any other area – to be part of the World Network of UNESCO Biosphere Reserves and, by virtue of its designed garden areas, also for inclusion on the UNESCO's World Heritage List.

Today's Garden Kingdom covers approximately 143 square kilometres between Dessau and Wörlitz and comprises the river landscape of the Elbe and the Mulde with wide, more extensively used flood plains and intensively used, diked old alluvial meadows.

Several landscape parks were incorporated in this "semi-natural" landscape dating from before 1800. Historical infrastructure elements, such as views, designed woody plantings, buildings, paths and small-scale architecture form to this day the main network of the total work of art that is the Dessau-Wörlitz Garden Kingdom. In line with the uniqueness of the landscape and the ideas of beauty of the day, single oaks, so-called solitary trees, were conserved or planted on the alluvial meadows. The dykes and ditch systems that made the land usable for man in the first place have been maintained and extended. Fruit trees were also part of the design to beautify the landscape by the wealth of their blossom and, at the same time, to encourage the planting of trees and the use of the fruits by the local population. Prince Franz of Anhalt-Dessau had educational programmes on agricultural and fruit growing subjects conducted in the Garden Kingdom, but also implemented the ideas of new garden and building arts.

Fig 1: Schloss Luisium Palace

Substantial parts of the Dessau-Wörlitz Garden Kingdom were recognised as a UNESCO Cultural World Heritage site in December 2000. It is considered to be an outstanding example of the implementation of the philosophical principles of the Enlightenment in a landscape design and of the harmonious combination of art, education and business.

Fig. 2: Warning Altar in Wörlitz Park: The oldest combined monument to nature conservation and culture in the Garden Kingdom. The altar bears the inscription „Wanderer, Achte Natur und Kunst und schone ihre Werke" [Walker, Respect Nature and Art and Protect Their Works].

In parallel, Dessau-Wörlitz Garden Kingdom was recognised as a historical cultural landscape by UNESCO as part of the Middle Elbe River Landscape Biosphere Reserve. This combination of two, complementary UNESCO programmes – unique in Germany – presents the opportunity to protect and conserve a "museum landscape" of international historical-cultural significance and, at the same time, to encourage sustainable regional development. In line with the MAB Programme's Seville Strategy of 1995, here, in particular, it is a matter of bringing "knowledge of the past to the needs of the future".

Biosphere Reserves and Cultural Landscapes

"Conventional" cultural landscapes in buffer zones and transition areas

The buffer zones and transition areas of the German biosphere reserves are largely made up of cultural landscapes that correspond to the "organically evolved landscapes" from the UNESCO World Heritage definition. Mostly, they are peripheral landscapes far away from conurbations, where a relatively small-scale structure (also in terms of ownership and usage patterns) and extremely high species diversity has been preserved.

In the buffer zones of the biosphere reserves in particular, the farms are often small, production strictly limited due to locational factors and the farms are no longer economic in spite of high agricultural subsidy levels. In the long term, this will probably be hard to reconcile with the principles of the UNESCO Man and the Biosphere Programme (MAB) (economic sustainability, transferability).
Contractual nature conservation and cultural landscape programmes are currently the best ways of conserving cultural landscape and creating income for the farmers. However, in the long term, alternatives must be found that are better integrated in agricultural and forestry operational processes (cf Chapter 3.3.2).
The fact that the very high level of biodiversity in the buffer zones of the German biosphere reserves and in comparable cultural landscape of their transition areas must be conserved is indisputable. But "sustainability" does not just mean ecological, but also economic and social viability. What economic and social services could such landscapes provide today if they are no longer essential for food and wood production?
One way out appears to be the production of especially high quality food and forest products. Their higher market revenues can compensate for the locational disadvantages of production. Today, there are already over 150 regional brands for agricultural produce in Germany (cf Chapter 4.1, 4.4 and 5.2). Certification is also increasingly well-received in forestry (cf Chapter 3.3.3).

Nature conservation as a "market service"

The nature that has developed in the central European cultural landscapes over the last few centuries was a "by-product" of nature utilisation, in particular of agriculture. But – unlike in earlier times – modern agriculture no longer creates biodiversity as a "side effect". Nature in cultural landscapes thus becomes a separate quality that has to be "generated" deliberately and with the aid of special techniques.
Nature conservation as a "product service" of agriculture and forestry? The frame conditions have never been better for an expansion of the agricultural "product range" of this kind. The importance of peripheral areas in particular for food production has been falling rapidly for decades, to the same extent to which its value as a recreational landscape has been growing. We therefore have to look into the question as to which "products" of cultural landscapes are actually wanted by society and are thus encouraged. The second pillar of EU agricultural policy offers additional opportunities.
But what would a "product nature" of this kind look like, how could "production services" be practicably paid for? Experience to date – not least from biosphere reserves – shows that the solution is not as easy as it may appear at first glance. Orders to satisfy nature conservation originally planned as a sovereign task (e.g. mowing a poor meadow as a consequence of contracts between the state and the owner) can initially lead to a thoroughly sensible result. But practice shows that farm structures remain unchanged (and centred on food production). The emotional identification with the new task remains low, a "market" does not emerge (cf Chapter 3.3.2).
If the "production" of nature is paid for according to performance in the normal farm structures that are familiar to the farmer, this also leads to incentives to adapt his own farm to this. And what's more: Nature is then a value that is in his own individual interest to conserve.
However logical this argument may be in theory, its realisation in practice will be difficult. What sort of nature should be "generated"? Who should decide this? How do we handle competing offers? How much is a certain type of nature or a corresponding service "worth"? The answers are simple if they are based on prescribed contractual services and on the losses resulting from reduced food production. But are these also the guiding conditions of the future?
Here, there is a wide area of development for biosphere reserves that is especially important for the MAB Programme: how can high-quality cultural landscapes that are important for nature conservation be conserved if food production is no longer to the fore and society's demands on these landscapes have changed in general? In the age of urbanisation and the "leisure society" this is a global problem. Solving it requires more efforts in all fields of "sustainability".

Information or emotions?

The basic attitude of the local population to "its" nature and landscape occupies a key position here. The development of cultural landscapes in the past was an interactive process between nature and the people who lived in it. It is only with technical emancipation that this became a human "intervention" and – as the word itself says – something not balanced, alien, mechanical, threatening. With the consequence that "communication" with nature, understanding its form, colour and uniqueness, has been lost. But is it really about an "understanding" of nature, i.e. about rational findings, or about "experience", "wonder", "joy", etc. and also about "pride" in living where the tree frog can still live, in other words is it about emotional values?
Our ancestors could not afford sentimentalities of this kind. Their lives – at least in rural areas – were always shaped by a form of nature that gave less than was needed for everyday life. But "respect" for nature remained in place.
Our "modern" societies categorically suppress thoughts of this kind. This is exactly where the work of the biosphere reserves must start. "Peripheral regions" are not "second class" areas. To a large extent they are those regions whose inhabitants have conserved them in such a way that it is emotionally worthwhile to be in them and live in them (cf Chapter 5.5).

Well over 10 million people seeking relaxation visit German biosphere reserves every year. And this is not to find out about a form of agriculture that is hopelessly old-fashioned and unprofitable (cf Chapter 3.3.4).

Literature on the Subject

FLADE, M., PLACHTER, H., HENNE, E. & K. ANDERS (Eds.) (2002): Naturschutz in der Agrarlandschaft. Ergebnisse des Schorfheide-Chorin-Projekts, Wiebelsheim.

FUKAREK, F. (1980): Über die Gefährdung der Flora der Nordbezirke der DDR. In: Phytocoenologia 7, pp. 174-182.

HABER, W. (1995): Concept, Origin and Meaning of 'Landscape'. In: VON DROSTE, B., PLACHTER, H. & M. RÖSSLER (Eds.): Cultural Landscapes of Universal Value. Components of a Global Strategy, Jena, pp. 38-41.

PHILLIPS, A. (1998): The Nature of Cultural Landscapes – a Nature Conservation Perspective. In: Landscape Res. 23, pp. 21-38.

PLACHTER, H. (1999): The Contributions of Cultural Landscapes to Nature Conservation. In: BUNDESDENKMALAMT (Ed.): Denkmal – Ensemble – Kulturlandschaft am Beispiel Wachau, Wien, pp. 93-115.

PLACHTER, H., BERNOTAT, D., MÜSSNER, R. & U. RIECKEN (Eds.) (2002): Entwicklung und Festlegung von Methodenstandards im Naturschutz. In: Schriftenreihe Naturschutz u. Landschaftspflege 70, Bonn.

PLACHTER, H., RÖSSLER, M. (1995): Cultural Landscapes: Reconnecting Culture and Nature. In: VON DROSTE, B., PLACHTER, H. & M. RÖSSLER (Eds.): Cultural Landscapes of Universal Value. Components of a Global Strategy, Jena, pp. 93-115.

SUKOPP, H., TREPL, L. (1987): Extinction and Naturalization of Plant Species as Related to Ecosystem Structure and Function. In: Ecol. Studies 51, pp. 245-276.

PUHLMANN, G., BRÄUER, G. (2001): Aufgaben und Ziele der Biosphärenreservatsverwaltung Mittlere Elbe zur Sicherung und Pflege des Dessau-Wörlitzer Gartenreichs. In: REICHHOFF, L., REFIOR, K. (Eds.): Schutz und Pflege historischer Kulturlandschaften als Aufgabe des Naturschutzes und der Denkmalpflege in Sachsen-Anhalt, Dessau, pp. 29-33.

PUHLMANN, G., JÄHRLING, K.-H. (2003): Erfahrungen mit "nachhaltigem Auenmanagement" im Biosphärenreservat "Flusslandschaft Mittlere Elbe". In: Natur und Landschaft 78, Heft 4, pp. 143-149.

3.2.4 Conserve Diversity! Practical Landscape Management

Josef Göppel

Until the 1980s the concept of "landscape management" in Germany was largely geared towards biotope conservation. The unification of Germany in 1990 resulted in a more comprehensive outlook: the focus of landscape management was no longer just small undeveloped areas, but also extensive cultural landscapes covering thousands of hectares.
At the same time, regional competition was flourishing and a close connection between intact landscapes and a general tendency to invest became apparent. Good opportunities for leisure and distinctive cultural characteristics were important locational factors. Over-exploited areas and those which had been neglected, however, fell behind in regional competition.
Turning attention to what was happening in the regional economy brought landscape management out of its niche existence. Building a network of natural living spaces was to be the main priority from then on and, alongside this, an increasing number of initiatives were taken to produce and market typical regional products.

All the village community is helping to provide a forest with a protective screen.

The Landscape Management Associations provided a new direction in terms of organisation. Above all, they quickly gained a foothold in areas with poor agricultural profits. Today, around 140 Landscape Management Associations operate in twelve of the 16 German *Länder*. Their key feature is the parity they encourage in decision-making between local politicians, the different occupations that work with the land and the executives of nature conservation groups. In practice, this fair and balanced approach creates a lot of trust and helps to settle conflicts on the spot.

Goals of the Landscape Management Associations

1. *To build an extensive network of natural living spaces to preserve the foundations of life in all cultural landscapes in Germany.*
2. *To provide agriculture with a reliable supplementary income from nature conservation and to support the marketing of typical regional products.*
3. *To provide initiatives for an ecologically-focused economic development and an environmentally friendly use of land. These should bring out the exceptional qualities and activate the strengths of the individual regions.*
4. *To open eyes for the landscape on their doorstep and allow all levels of the population to experience real nature through targeted actions.*

Source: German Landscape Management Association (DVL) / www.lpv.de

The groups that work with the land include farmers and foresters, hunters, fishermen, tour operators, typical regional tradesmen and also people pursuing leisure activities, such as climbing or white water rafting. The motto of the Landscape Management Associations is to include all good ideas and exclude no one.

As a rule, the private associations are welcome helpers for the government authorities. Like the associations for voluntary social work, Landscape Management Associations help to carry out governmental tasks.

Fifteen years of experience show that considerably more initiatives can be implemented through such action alliances than by the government administration working alone. Landscape Management Associations do not draw up plans. Their responsibility is to prepare, run, control and render the accounts of concrete projects. Plans are not in short supply in Germany – the bottleneck occurs when they are being implemented!

Besides building a network of habitats for creatures living alongside humans and strengthening the regional added value, a permanent task of landscape management is to introduce people to the nature on their doorstep.

Many people are so wrapped up in the civilised world today that they almost fail to notice their surrounding landscape. The average German spends less than one hour a day in the open air. Television and the internet lead children to believe in a world in which everything seems possible.

However, even the best films can not replace the smell of a meadow in summer, the coolness of a shady running stream or an apple picked directly from the tree. If adults haven't experienced real nature as children, it will be difficult to inspire them with enthusiasm about its conservation. This demonstrates the broad and rewarding areas of responsibility confronting the Landscape Management Associations.

Practical landscape management faces particular challenges in biosphere reserves. On the one hand, more accurate and specialist material is available and as a rule, there are better principles for planning. On the other hand, action taken must satisfy higher quality standards.

There are also differences in the working. The management of a Landscape Management Association normally is left on its own. However, when cooperating with a biosphere reserve a specialist, qualified administration is working by its side. This leads to more intensive preparation and follow-up controls of all stages of work. In short, specialist landscape management expertise within biosphere reserves is, for the most part, greater than that outside. New incentives for the whole country often result from this.

The financing of nature conservation and landscape management is a task of prime importance for the *Länder* in Germany. In comparison, all other political matters only fulfil a complementary function. Unfortunately, during times when finances are limited, commitment to sustainable development can wane. With the exception of the *Länder* Bavaria and North Rhine-Westphalia, the Landscape Management Associations and Biological Centres still do not have a guaranteed financial basis.

In view of the uncertain professional financial future of many managers, their idealism is admirable. It is politically fair,

Special Report of the Federal Government's Advisory Council of Experts on the Environment in 1996:

Concepts for a Sustainable Use of the Rural Environment

"Landscape Management Associations have proved to be effective organisations for putting the goals of landscape management and nature conservation into practice. (...) Integrating all groups affected has been a successful way of promoting acceptance and taking advantage of the wealth of experience of all participants. (...) The German Advisory Council of Experts on the Environment recommends that Landscape Management Associations be institutionalised and supported in the implementation of regional land use concepts and communal landscape planning."

however, to honour payment for the services provided by private associations that fulfil public duties.
For many men and women involved in the landscape and regional initiatives the political background must also be mentioned. They feel the pressure coming from globalisation to standardise all forms of life, language, clothing, food, building styles and leisure activities. People who have set themselves the goal of preserving the diversity of all living creatures are affected by this to the core. The ecological focus of the long-standing goal of landscape management to conserve diversity is suddenly taking on a cultural dimension. Is the balance between regional concerns and global economies successful?
The unlimited liberalisation brought about by world trade is increasingly criticised because it has broadened, not narrowed the gap between the rich and the poor. Regional initiatives counteract this to a certain extent. As a result of regional initiatives, people settle in well-defined cycles of life with an independent profile and thus contrast the people who live in the centralised industrialised world and are being deprived of their soul.

Josef Göppel

Mowing at a steep slope is hard work.

Literature on the Subject

ARBEITSGEMEINSCHAFT BÄUERLICHE LANDWIRTSCHAFT (Ed.) (1997): Leitfaden zur Regionalentwicklung, Rheda-Wiedenbrück.
DVL (DEUTSCHER VERBAND FÜR LANDSCHAFTSPFLEGE e. V.) (1998): Waldrand – Hinweise zur Biotop- und Landschaftspflege, Beutel.
DVL (DEUTSCHER VERBAND FÜR LANDSCHAFTSPFLEGE e. V.) (2001): Fledermäuse im Wald – Informationen und Empfehlungen für den Waldbewirtschafter. In: Landschaft als Lebensraum, Ausgabe 4, Ansbach.
DVL (DEUTSCHER VERBAND FÜR LANDSCHAFTSPFLEGE e. V.) (2002): Erprobungs- und Entwicklungsvorhaben: Reptilienlebensraum Lechtal (Voruntersuchung), Ansbach. Unpublished.
FRANCÉ, R. (1923): Die Entdeckung der Heimat, Asendorf.
GÖPPEL, J. (1993): Landschaft als Lebensraum. Issue 1 of the DVL document series.
GÖPPEL, J. (1996): Denken in Landschaften. Cadolzburg: ars vivendi.
GÖPPEL, J. (1999): Regionen im Aufbruch. Issue 2 of the DVL document series.
GÖPPEL, J. (2000): Die Farben der Zukunft. – Wie regionales Wirtschaften erfolgreich wird. Issue 5 of the DVL document series.
GÜTHLER, W. (1999): Landschaftspflegeverbände – Bündnisse für die Natur. In: KONOLD, W., BÖCKER, R. & U. HAMPICKE: Handbuch Naturschutz und Landschaftspflege, Landsberg.
GÜTHLER, W. (2001): Agrarumweltprogramme als Perspektive der Kooperation zwischen Landwirtschaft und Naturschutz. In: OSTERBURG, B., NIEBERG, H. (Eds.) (2002): Conference Publication "Agrarumweltprogramme" – Konzepte, Entwicklungen, künftige Ausgestaltung", Conference on 27 – 28 November 2000 in Braunschweig. Landbauforschung Völkenrode Sonderausgabe 231.
GÜTHLER, W. (2002): Zwischen Blumenwiese und Fichtendickung: Naturschutz und Erstaufforstung. In: BUNDESAMT FÜR NATURSCHUTZ (Ed.): Schriftenreihe angewandte Landschaftsökologie, Ausgabe 45, 133, Bonn.
GÜTHLER, W., KRETZSCHMAR, C. & D. PASCH (2003): Verwaltungsprobleme des Vertragsnaturschutzes und mögliche Lösungsansätze. In: BUNDESAMT FÜR NATURSCHUTZ (Ed.): BfN-Skripten, Ausgabe 86, Bonn.
HAHNE, U. (1984): Endogene Entwicklung, Theoretische Begründung und Strategiediskussion. Arbeitsmaterial der Akademie für Raumforschung und Landesplanung, Nr. 76, Hannover.
MAIER, J. (1997): Nachhaltige Regionalentwicklung und regionale Energieversorgung. In: Das Prinzip der Nachhaltigen Entwicklung in der räumlichen Planung. Arbeitsmaterial der Akademie für Raumforschung und Landesplanung, Nr. 238, Hannover, pp. 138-141.
MAIER, J, TROEGER, G. & J. WEBER (1985): Regionale Selbstverwiklichung im Tourismus, Darstellung und Kritik des Konzeptes einer endogenen Regionalentwicklung am Beispiel peripherer Mittelgebirgslagen. In: Informationen zur Raumentwicklung, H. 1 1985, pp. 21-33.
MAIER, J. (1997): Anforderungen an die Umsetzung des Gebots der Nachhaltigkeit in der ländlichen Entwicklung aus der Sicht des Regionalwirtschaftlers, Conference Publication by the Bayerischen Akademie Ländlicher Raum, Memmingen.
MOSE, I. (1993): Eigenständige Regionalentwicklung – Neue Chancen für die ländliche Peripherie? In: Studien zur angewandten Geographie und Regionalwissenschaft, Band 8, Vechta.
TSCHUNKO, S., GÜTHLER, W. (1997): Landschaftspflegeverbände in Bayern – Erfahrungen und Perspektiven. Unpublished study commissioned by the Bavarian State Ministry for Regional Development and Environmental Affairs.
WALDERT, H. (1992): Gründungen – Starke Projekte in schwachen Regionen, Wien.

3.2.5 *The Importance of Nature Rangers*

Beate Blahy and Gertrud Hein

The Criteria for Designation and Evaluation of UNESCO Biosphere Reserves in Germany specify that full-time administration, including nature rangers must be safeguarded (National Criteria cf Annex, p. 164).

In the 14 German biosphere reserves there are now approximately 200 full-time employed nature rangers. In the areas that also comprise national parks nature rangers work both for the national park and the biosphere reserve.

Tasks of Nature Rangers

Environmental education and public relations

For most visitors as well as for the local population, very personal contact with the people who work in the biosphere reserve administration is very important. In the biosphere reserves it is mainly the nature rangers who perform this task because they meet large numbers of people during excursions or conversations at the information stands.

Personal encounters of this kind are essential in shaping the public image of biosphere reserves.

For example, environmental education and public relations work includes:

- excursions for various target groups,
- guided tours and looking after visitors in information centres,
- information stands at events,
- helping to arrange campaign days in the protected area and
- contacts for the players in the protected area from the fields of politics, administration, associations and the public.

Beate Blahy

At the information stand

Taking care of project weeks with school classes on various environmental themes are part of the work of nature rangers, as is the management and mentoring of children and youth groups from the region over many years. This makes it possible to teach children about environmental and nature conservation issues, ideas on sustainability and Agenda 21 in a way that is appropriate for their ages and over a long period of time.

Furthermore, there are other educational offers for different age and social groups, where cooperation with other educational institutions and associations has proved valuable.

Controls

Regular controls of the area, at shorter intervals in the visitor season, at the weekends and in areas used extensively by tourists, help to ensure that the existing regulations for the protected area are voluntarily followed or implemented.

The presence of nature rangers means that violations of the path regulations or administrative offences, such as camping in a nature conservation area, lighting fires, letting dogs run free, digging up protected plants or disturbing protected animals are greatly restricted or prohibited. In most of the *Länder* nature rangers do not have sovereign powers and depend on close cooperation with the law enforcement authorities.

In the protected areas in particular, i.e. in the core areas and buffer zones of a biosphere reserve, the mere presence of nature rangers is motivation enough for many visitors to adhere to the rules and bans.

Beate Blahy

Nature rangers are tasked with a considerable part of the work done to conserve endangered species.

Biotope and species protection, monitoring, support academic work

Nature rangers do a large proportion of the work in protection programmes for endangered animal and plant species. This includes recording breeding and rearing, e.g. of White Storks (*Ciconia ciconia*), conducting regular bird surveys and recording data on individual species (amphibians, reptiles,

birds). In long-term studies, such as the Integrated Environmental Monitoring, they regularly collect certain data, thus helping scientists.
Within the context of protection, the control and development of designated Fauna-Flora-Habitat areas, nature rangers perform mapping and control tasks.

Landscape management measures, contractual nature conservation

Nature rangers conduct diverse landscape management measures and give advise to the land users. Important tasks include administering the contracts that are concluded within the context of contractual nature conservation. Nature rangers seek appropriate areas for the contracts, advise the contractors and monitor implementation of the agreed measures.

Technical Foundations of the Work

Qualified, well-trained employees are needed to implement nature conservation measures and for expert visitor care. That is why numerous nature rangers have already completed the course to become a "Qualified Nature and Landscape Manager".
In Germany this further training course leads to the first non-academic vocational qualification in nature conservation. It has been available across the country since 1998.
The general material plan recommendation comprises 640 hours (= 16 weeks) with the following four blocks of subjects (BMELF 1999):

- Foundations of nature conservation and landscape management
- Information activity and visitor care
- Measures for nature conservation and landscape management
- Economics, law and social affairs

Controlling protected areas is only one of the many tasks of nature rangers.

At the end of the course there is an examination (practical, written, oral). The requirement for admission to the exam is a completed vocational training in a "green" profession and proof of many years of professional experience. This further training course teaches planning and practical landscape management as well as skills in the fields of visitor information, protected area management and communication.
Continuous further training of nature rangers will be necessary to ensure that nature rangers will continue to work successfully in biosphere reserves in the future.
Their work is important for the biosphere reserves: due to their many contacts with visitors and locals they help in continuously and expertly communicating the idea and objectives of the UNESCO Man and the Biosphere Programme (MAB) in the region.

Regular controls by nature rangers help to protect nature conservation areas.

Literature

BUNDESMINISTERIUM FÜR ERNÄHRUNG, LANDWIRTSHCAFT UND FORSTEN (BMELF) (1999): Rahmenstoffplan für die Durchführung von Fortbildungslehrgängen zur Vorbereitung auf die Prüfung zum anerkannten Abschluß "Geprüfter Natur- und Landschaftspfleger", Bonn.
BUNDESVERBAND BERUFLICHER NATURSCHUTZ e. V. (2000): Geprüfte/r Natur- und Landschaftspfleger/in, Bonn.

3.3 *Sustainable Regional Development*

3.3.1 *Sustainable Economic Development*

Werner Schulz

Why and Towards What?

"Sustainability" – the task for the future – is a multifaceted subject – the range of literature is vast (e.g. cf BUNDESREGIERUNG 1999). In the opinion of the German Council for Sustainable Development (*Rat für Nachhaltige Entwicklung*) appointed by the Federal Government, sustainable development aims at a future in an ever bigger and brighter world, with a clean and healthy environment and with natural diversity remaining intact, a world in which there is more democracy and prosperity and where the shared cultural heritage is maintained (www.nachhaltigkeitsrat.de). Living and working in a way that is not at the expense of future generations or people in other parts of the world – that is the key principle of sustainability.

Sustainability is much older than the current popularity of the term would lead us to believe. In fact, the history of sustainability goes back to Saxony in the baroque age. In Freiberg around 1700, Chief Inspector of Mines Carl von Carlowitz developed a counter model to the severe degradation of forests practised until then: to conserve the forest resources in the long term, he recommended that only so much wood should be felled as could grow back through reforestation.

Around 300 years later, the predatory exploitation of the natural foundations of life has not diminished, but much rather attained global dimensions. And thus, in 1987 the Brundtland Commission took up the principle of sustainability and described sustainable development as a system of management that "satisfies present needs without compromising the ability of future generations to meet their own needs".

At the UN Conference on Environment and Development in Rio de Janeiro in 1992 the United Nations declared sustainability to be the guiding principle for the 21st century. Ever since, the guiding principle of sustainable development has gained a foothold in political institutions and programmes at all levels.

In Rio, industrialised and developing countries agreed on the confirmation of the future goal of global, sustainable development. Since the "Rio plus ten" follow-up conference in Johannesburg in 2002 at the latest, this goal has been defined so that it goes beyond the mere maintenance of the ability of the ecological system to function. Much rather, the idea includes the goal of a life with human dignity based on individual personal development, both for the current and future generations, encompassing social, ethical and economic dimensions.

The European Union, too, made sustainable development into a central component of its common policy in the 1997 Amsterdam Treaty. At the 2001 Gothenburg Summit, it presented a strategy entitled "A Sustainable Europe for a Better World" that expanded the strategic goals for economic and social policy that had been laid down in Lisbon one year earlier with an ecological dimension.

The implementation of the European objective at national level defines the Federal Government's sustainability strategy under the title "Perspectives for Germany". In this, the Federal Government defines sustainability as an interdisciplinary task that is to be a fundamental principle in its policy in all fields in future. On the whole, the strategy formulates guiding principles of sustainable action for the key areas of energy, transport, health protection and food, family and old people, education and innovation.

Buzzword
"Sustainable Economic Development"

The sustainability approach aims at bringing together economic performance, social responsibility and environmental protection to facilitate fair development opportunities for all countries and to preserve the natural foundations of life for future generations. Currently, throughout the world there are around 70 attempts to bring this guiding principle ("regulative idea") closer to operationalisation. Examples:

- *If the ecologists have their way the ecosystems should not be overloaded by a use of the resources there.*
- *Most economists view sustainable development as an economic form that has to ensure that the same welfare will be available for future generations as for those of today.*
- *Physicists call for the conservation of biological systems that are stable within themselves, and chemists would like all anthropogenically influenced substance cycles to be closed where possible (i.e. "recycling").*

Particularly drastic examples of non-sustainable economic development are

- *deforestation in the Mediterranean region by the Romans and the destruction of tropical forests today,*

- overfishing of the oceans by ever more perfect catching techniques and
- the steppisation of large parts of the former Lake Aral in Russia as a consequence of the diversion of large quantities of water to irrigate agriculture.
Examples of sustainable economic development are harder to find, especially if not all forms of economic activity that owe their permanence only to the low levels of technical intervention in the past are to be called sustainable. In principle, the following types of economic activity can be considered sustainable:
- the cultivation of centuries-old rice terraces in China and Indonesia,
- various forms of agricultural forest use in Africa and Latin America, e.g. agro-forestry and
- the cultivation of Alpine pastures from the 17th century to the end of the Second World War.

Numerous German local authorities together with several thousand cities and communities throughout the world are on the way towards a local agenda. The trigger for this movement was the final Rio document of 1992: Agenda 21. This global programme of action for sustainable development was signed with binding effect by most countries on earth – including Germany. By now, agenda processes referring to individual towns and cities have been set in motion in practically all German cities.

Criteria for Sustainable Economic Development

Sustainability in companies

Various fields of action show how the model of sustainability can be specifically implemented in companies – for example in the areas of agriculture, tourism, construction, transport, retail and skilled crafts:

■ **"Ecology"**

- Sensitive dealings with resources (input)

Examples: reduced consumption of raw materials, process materials in tonnes; growing proportion of regenerative fuels in per cent; proportion of materials grown or produced according to environmental criteria (e.g. organic farming, "Öko-Tex-Standard 100").

- Reducing environmental pollution from substance inputs (output)

Examples: reducing the specific waste volume; reducing the amount of special waste; increasing the recovery rate

- Responsible handling of ecosystems

Examples: does the company respect the protection and conservation of species diversity, natural spaces and ecosystem when extracting, using or processing resources (e.g. soil, fish stocks) at its own premises and at the supplier's premises? Does the company create additional valuable areas in terms of nature conservation, or at least green spaces (e.g. by planting, unsealing, growing native species or harmonisation measures above and beyond what the law demands)? Does the company execute building projects (new buildings or extensions) in such a way as to save land and to avoid overdevelopment (e.g. within the municipality, using brownfield sites, using existing buildings, optimising land use)? Are biological, aesthetic and humane principles considered in building projects (e.g. fitting the buildings into the local environment, interior design conducive to human well-being)?

- Minimising the risks for people and the environment

Examples: reducing the proportion of hazardous substances; number of environmentally relevant accidents and incidents that have to be reported in the last five years

- Environmentally sound products and procedures

Example: proportion of production processes in per cent that have been assessed according to ecological criteria in the last five years

- Global ecological responsibility

Example: does the company sell waste for disposal or recovery to other countries, in particular developing or newly industrialising countries?

■ **"Social Issues"**

- Jobs, training and employees' interests

Examples: does the company pursue the long-term creation and safeguarding of jobs that are as permanent as possible? Does it offer skilled part-time jobs? Does it offer training places and opportunities for further training? Does the company allow a staff council and trade union activities without restrictions?

- Health and safety at work

Examples: industrial accidents, occupational diseases and days off due to sickness should be reduced as far as possible; jobs should be designed according to ergonomic criteria as far as possible; the company should offer company sport.

- Equality of the sexes

Examples: the company should ensure that there is a high proportion of women at managerial level; measures to encourage women (e.g. for returning to work) should be offered.

- Social consideration

Examples: does the company employ the disabled at least in line with the statutory quota? Does the company consider the cultural needs of foreign employees?

■ **"Economy"**

- Long-term company security

Example relevant to ecology: what costs/savings were linked to company environmental protection measures?
- Wealth creation and equitable distribution

Example relevant to ecology: what percentage of profits is accounted for by funds for eco-sponsoring?
- Orientation to needs

Example relevant to ecology: what percentage of products are ecologically dubious (e.g. disposable products)?
- Regional/global responsibility

Examples relevant to ecology: what percentage of goods and services is regional? What is the percentage of material for which fair prices are paid to developing and newly industrialising countries in particular (cf e.g. Eco-Fair logo)?

Sustainable state

In 1995 the G7 Environment Ministers decided on the "ecologisation" of their activities at their meeting in Hamilton in Canada. Environmentally friendly public procurement was viewed as a main contribution to the "greening" of government. (UMWELTBUNDESAMT 1999; BUNDESUMWELTMINISTERIUM/UMWELTBUNDESAMT 2001)

At the Eco-Procura Conference in Lyons in the year 2000, leading personalities from local government asked governments at all levels to contribute in a credible way to the process of sustainable development by "ecologising" their policy and activities. Sustainable development requires that all sections of our society work on its implementation. Nevertheless, the public sector has a key role because it can take on the pacemaker function with its role as a model. In Lyons, the following criteria for sustainability were laid down for the state level:

- consideration of the critical loads of the environment by means of political management systems, such as management of the natural balance and by setting reduction targets for emission relevant to climate, nitrogen emission, water consumption, waste generation and land consumption;
- enabling employees at all levels of the administration to implement and continuously improve the "ecologisation" of the administration by means of appropriate training and by environment and energy management systems (cf BUNDESUMWELTMINISTERIUM/UMWELTBUNDESAMT 2001).

Sustainable consumption

According to the Federal Environmental Agency, up to 40 per cent of the environmental pollution is due to human consumption patterns. That is why changing consumption patterns is an important point in the Programme of Action for Agenda 21. However, not only the individual consumer is responsible for sustainable consumption. This can be seen especially clearly in product-related environmental protection. Here, the manufacturing industry, skilled crafts and retail are faced with a challenge. They have to offer environmentally friendly products and services. Only if consumers know of more environmentally friendly alternatives to conventional products will they decide to buy them. But sustainable consumption is more than buying environmentally sound products. It is also about changing by means of new consumption styles and wealth orientation.

Corporate Practice

Participation in EMAS and ISO 14001

The EC EMAS (Eco-Management and Audit Scheme) is based on European Council Regulation 1836/93/EEC of 29 June 1993 allowing voluntary participation by companies in the industrial sector in a Community eco-management and audit scheme.

The aim of the regulation, which has been in force since 1995, is to achieve a uniform system for the continuous improvement of company environmental protection throughout Europe. The regulation was revised in 2001. On a voluntary basis the companies should be encouraged to develop environmental programmes and environmental management systems, conduct audits and draw up environmental declarations. In the member states almost 3,800 company locations have been registered since the entry into force of the EC EMAS regulation, between autumn 1995 and mid-2003. Around 65 per cent of the total participating in EMAS is currently in Germany. Austria, Spain and Sweden follow in places two, three and four.

Alongside the European EMAS, ISO standard 14001 is a central instrument for setting up environmental management systems, in which almost 50,000 companies are currently participating (cf Fig. 1). Germany is currently one of the main participating countries in this system with just under 4,000 certified companies.

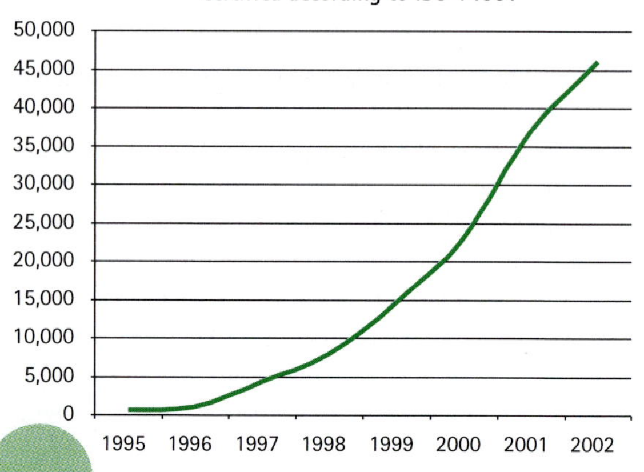

Fig. 1: Participation in the ISO 14001 system (as of: end of 2002)
Source: The ISO Survey of ISO 9000 and ISO 14000 Certificates, ISO 2002.

Experience, trends and potential

The team from the oekoradar joint project (www.oekoradar.de) commissioned the ifo Institute for Economic Research Munich (*Institut für Wirtschaftsforschung*) with a study on the state of development (ÖKORADAR 2002). In total, 5,788 companies – mainly the managing directors – were asked about the extent to which the subject of sustainable economic development had gained a footing in operational practice.

In response to the question of the subject of sustainability, over one third of the companies covered by this study stated that they had already tried to implement the guiding principle into specific targets or draw up measures for this several times. A quarter of those questioned were thinking of classical environmental protection.

Although the majority of companies in Germany still behaves largely passively, around 58 per cent of those questioned assume that the future will demand more commitment of them and that the importance of social and ecological responsibility will increase (cf Fig. 2).

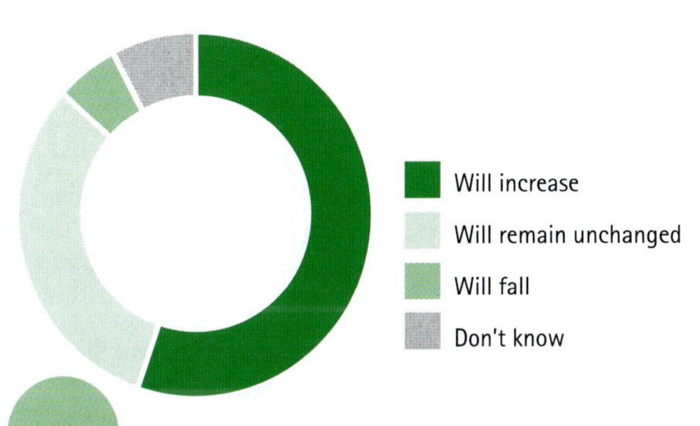

Fig. 2: Assessment of companies' social and ecological responsibility (in per cent)

Literature

BUNDESREGIERUNG (1999): Stichwort "Nachhaltigkeit", Bonn.

BUNDESUMWELTMINISTERIUM/UMWELTBUNDESAMT (Ed.) (2001): Handbuch Umweltcontrolling. 2. Auflage, München.

ÖKORADAR (2002): Nachhaltiges Wirtschaften in Deutschland. Erfahrungen, Trends und Potenziale, München.

UMWELTBUNDESAMT (Ed.) (1999): Handbuch Umweltfreundliche Beschaffung. 4. Auflage, München.

3.3.2 Sustainable Land Management

Jürgen Rimpau

The cultural landscapes in the German biosphere reserves have been shaped by agricultural and forestry activities. In Germany in the past the focus was on the protection of particularly valuable natural commodities. In future, however, it will be decisive to put sustainable development as a whole on top of the agenda. As well as the special ecological protective function, social and economic goals of regional development have to be pursued in the further development of biosphere reserves. In this connection, not only the income of farms has to be secured, the farms must also be capable of meeting the special challenges arising from globalisation, World Trade Organisation (WTO) negotiations, European Union (EU) agricultural reform and consumer demands.

Some marginal sites (medium mountain locations with high proportions of clay; sand without precipitation) will be taken out of production. Strategies must be developed for the fate of land becoming fallow (keeping it open as fallow land; natural succession; active rededication to semi-natural land; reforestation). The spectrum of species in these parts of the landscape will thus change in the long run and permanently.

Agriculture is responding to these challenges with the following strategic approaches:

- **Promoting multifunctionality:** Agriculture not only produces food and fodder, in future it will also produce regenerative raw materials and climate-neutral energy. At the same time, the farms are diversifying into direct marketing by further processing their raw products into premium products themselves, into catering, farm holidays and cultural events. In this connection, the building substance is often rededicated and, thus, conserved. Natural monuments are revitalised. Biosphere reserves can be model regions for the development of marketing structures and for interlinking within agriculture. At the same time, new forms of regional development ("development team") are to be tested.
- **Market for environmental services:** Nature and environment have a high value for society and agriculture. This value is not countered by any activity on the market and no market price. Therefore, agricultural environmental programmes have been developed that demand and recompense defined environmental services. The farmer acts as a provider of environmental services. This replacement market must be developed nationwide. Open questions of where the measures are targeted and of evaluation could be tackled in biosphere reserves to set an example. To increase the attractiveness of agricultural environmental programmes, it is important to simplify the organisational structures and to increase

transparency. In this case, too, biosphere reserves could become a trial field and, where appropriate, take on a model character. Environmental services that are to be recompensed should be distinguished from those environmental services in agriculture that automatically arise as tie-in products in the production of foodstuffs. For example, these include keeping the landscape open, maintaining paths and drainage, caring for hedges and trees. On agricultural land these are all elements of "good technical practice". Although the definitions of "good technical practice" are laid down in agricultural legislation, they should be defined and further developed as far as possible within the context of scientific limits. This is necessary to lay down a clear distinction to the agricultural environmental services that are to be recompensed. There are scientific approaches for defining "good technical practice" to the requirements of the environmental protection goals. Biosphere reserves are the ideal location for testing whether these proposals are ready for practice. It is important that the rules of "good technical practice" also have to be in line with the requirements of sustainability goals.

- **Guiding principle of sustainable development:** A new guiding principle of "Sustainable Land Management" has been developed in agriculture. Starting with the familiar three pillars of sustainability (ecological, social, economic), the guiding principle refers to the multifunctionality of agriculture, which defines a whole array of sustainability goals. Due to the fundamental equality of the three pillars of sustainability, the sustainability goals are linked to each other. For this reason, individual goals of sustainability should not be highlighted or even dealt with one after the other. Only the carefully harmonised set of criteria should be tackled jointly and at the same time. Sustainability goals can compete with each other; furthermore they are not of equal weight. That is why the goals must be weighted and evaluated as they manifest very different functions and dimensions (for example, carbon dioxide pollution of the atmosphere and securing income for people in rural areas). This can best be achieved via a monetary evaluation. However, not all criteria can be assessed in terms of quantity (e.g. the aesthetics and diversity of landscapes). Precisely this property requires that there goals, which have a high value for society but no price due to the lack of a market, must not be undermined, but have to be described verbally or by means of alternative values and introduced into the decision-making processes. The key instrument of sustainability on the practical level is an integrated management system comprising classical controlling, business economics, quality management, environmental management, documentation and a seal of quality. In a horizontal comparison of farms, these figures (economic and ecological indicators) are reviewed annually. Analyses of weak points and corrections to courses will now be possible at short notice and permanently. Discussion is part of the binding element of sustainable farm management. The consumer's confidence in food can only be achieved if the links in the value added chain are jointly subject to the criteria of sustainability and present themselves to the consumer with coherent, convincing and transparent information on the quality and safety of food.

Literature on the Subject

KAGERBAUER, A. (2003): Regionalentwicklung und soziale Netzwerke. Master's thesis, Göttingen.

DLG-WWF (2003): Die Agrarumweltprogramme. Ansätze zu ihrer Weiterentwicklung.

RIMPAU, J. (2003): Nachhaltigkeit – ein neues Leitbild für die Landwirtschaft. In: Schriftenreihe der DBU.

RIMPAU, J. (2003): Die Anforderungen steigen – was tut die Landwirtschaft? Beispiele aus der Praxis: Pflanzenproduktion. In: 6. aid-Forum, aid-Schriftenreihe.

3.3.3 Sustainable Forestry

Hermann Graf Hatzfeldt

Forests are the last large more-or-less intact terrestrial ecosystem in Germany. They are of irreplaceable importance for species diversity, soil protection, water resources and the climate, not forgetting recreation and the national culture. Depending on the character of the biosphere reserve, forestry management is of greater or lesser importance for sustainable regional development.

The type of management is decisive for the contribution that forestry makes to sustainable development. Since the Brundtland Report of 1987 and the 1992 UN Conference for Environment and Development (UNCED) in Rio de Janeiro the new guiding principle has been "proximity to nature". Semi-natural forestry is based on the natural forest communities in the site in question, as they would probably have grown without human intervention. We therefore need unmanaged protected areas not influenced by people, where the natural processes and mechanisms of self-regulation that take place in the forest ecosystem can be observed. This is a major concern for designating core areas.

Franz Straubinger

Natural forestry: multi-layered mixed forest, in which the forest renews itself under the canopy of old trees

Franz Straubinger

Typical harvested age-class forest

In the transition areas and – with limitations – also in the buffer zones of the biosphere reserves, the demanding task is to use these processes and mechanisms economically without damaging them. In other words, to work with nature instead of working against it. The previously usual separation of regeneration, maintenance and use has ceased – both in terms of space and time. Every felled tree is also used to care for its neighbours and to regenerate the forest. Interventions are limited to the absolutely necessary minimum in this way. In comparison to traditional management, the semi-natural version is attractive in terms of business management: expenditure falls, revenue rises and risks are avoided. Ecological management therefore not only benefits the forest, but is also economically worthwhile. By contrast, non-ecological management can turn out to be expensive.

Semi-natural forestry is a model for sustainability when dealing with nature, which is also applicable to other sectors of the economy. It vividly demonstrates that conservation and use do not have to be a contradiction in terms. Protection and use can be profitably combined in the forest if people manage nature cleverly and treat it appropriately. Claiming a model character, the German biosphere reserves have appropriated the principles and experience of semi-natural forestry and have benefited from them. In the buffer zones and transition areas the forest should be managed in a semi-natural way where possible, in core areas it is left to its own devices.

However, it is a long way from the idea to its realisation. New concepts cannot be implemented in practice overnight. One difficulty is that a new ecological direction for forestry requires a radical change in how foresters see themselves. To really manage forests "naturally" foresters have to understand the forest as a complex ecosystem whose dynamic processes do not have to be mastered, but understood, gently guided and imitated parasitically at the same time. Master foresters must become forest partners. It is hardly surprising that this learning process is arduous and drawn out.

A complicating factor is that semi-natural forestry absolutely requires forest-compatible stocks of hoofed game. If the game densities are higher, the principles of naturalness cannot be realised on a large scale. However, the current forest game situation in Germany is diametrically opposed to the implementation of the "semi-natural forestry" model –

Franz Straubinger

Multi-layered mixed forest with heavy wood production; single trees are harvested.

regardless of the type of ownership. In some German biosphere reserves the pressure of game in the buffer zone and transition areas is so great that there has to be hunting in the core areas so that the natural development of the forest is not jeopardised.

What should be done? How can the naturally managing of the forests in German biosphere reserves be advanced in the future so that their potential for sustainable development is fully realised? An innovative new instrument appears to be ideally suited to this: forest and wood certification, especially under the system of the "Forest Stewardship Council (FSC)" (www.fsc-deutschland.de, www.fscoax.org).

FSC Trademark© 1996 Forest Stewardship Council A.C.

The aim of certification is to protect and promote comprehensive sustainable forest development. Defined minimum requirements and regular inspection of compliance will give concrete instructions to improve the sustainable management of the forest. Furthermore, market economy incentives will promote the use of wood from sustainable sources. In view of the globalisation of the wood markets and the suppression of regional production due to imports from disputed sources this aspect will become increasingly important. Last but not least, the system of certification is an ideal teaching and learning model. The term "sustainability" can be defined in its various economic, social and ecological facets using the example of forestry and achievements can be communicated to the public with great effect.

The German MAB National Committee recommends that the management of the forests sited in biosphere reserves should be certified according to the demanding standards of the FSC Working Group in Germany. If the highly forested biosphere reserves follow the recommendation, semi-natural forestry will be implemented in an exemplary fashion in the German biosphere reserves within the comprehensive meaning of Rio. This would not only meet the model requirements of biosphere reserves, but also benefit the state of forests in biosphere reserves and regional development.

Literature on the Subject

Arbeitsgemeinschaft Naturgemäße Waldwirtschaft (ANW) (Ed.)(1999): Der Dauerwald: Zeitschrift für naturgemäße Waldwirtschaft. Edition 20, Butzbach/Nieder-Weisel.

Bode, W. (Ed.) (1997): Naturnahe Waldwirtschaft: Prozessschutz oder biologische Nachhaltigkeit? Holm: Deukalion.

Dieter, M., Thoroe, C. (2003): Forst- und Holzwirtschaft in der Bundesrepublik Deutschland nach der neuen Sektorenabgrenzung. In: Forstwirtschaftliches Centralblatt 122. Berlin: Blackwell Verlag.

Hatzfeldt, Graf H. (Ed.)(1996): Ökologische Waldwirtschaft: Grundlagen – Aspekte – Beispiele. Heidelberg: C. F. Müller.

Kassel, R., Bücking, M.; Roeder, A. & M. Jochum (2003): Ergebnisse der waldbaulichen Gutachten in Rheinland-Pfalz auf Landesebene.

Klins, U. (2000): Die Zertifizierung von Wald und Holzprodukten in Deutschland – eine forstpolitische Analyse. Dissertation der forstwissenschaflichen Fakultät der Technischen Universität München.

Meidinger, E., Elliot, C. & G. Oesten (Eds.) (2003): Social and Political Dimensions of Forest Certification. www.forstbuch.de, Remagen-Oberwinter.

3.3.4 Sustainable Tourism Development

Barbara Engels and Beate Job-Hoben

Introduction

The 14 German biosphere reserves have everything in terms of nature and countryside that a holiday-maker could desire: the biggest contiguous mudflat landscape on earth in the north, high mountains in the south, as well as coasts, rivers and forests.

They are representative of natural and cultural landscapes that are unique in their model character. And it is precisely this that makes the biosphere reserves attractive destinations for a holiday in Germany based on nature and the countryside.

What Are the Destinations? – Trends in Tourism

The development in tourism in Germany is characterised by different key areas: in addition to the increased interest in destinations in Germany (F.U.R. 2003), health and well-being tourism is also growing in popularity (F.U.R. 2002).

"Experiencing nature" and "purer air, cleaner water, getting out of the polluted environment" are among the most important holiday motives of Germans and can support the decision for a visit to a large protected area (PETERMANN, T., REVERMANN, C. 2002: 48). Moreover, it is precisely large-scale protected areas, i.e. nature parks, national parks and biosphere reserves, that offer a perfect "mixture of activity and calm", which is what many holiday-makers want (F.U.R. 2001).

The importance of large-scale protected areas for tourism in Germany can be seen in the growing numbers of holiday-makers. In the first half of 2003 the Hainich National Park recorded an increase in visitors of 20 per cent in comparison to the previous year, the proportion of outside visitors rose from 15 per cent to 30 per cent (NEWSLETTER FAHRTZIEL NATUR 19/03). The Bavarian Forest Biosphere Reserve/National Park registers around two million visitors per year, and the numbers of overnight stays in the surrounding municipalities have trebled since the National Park was established in 1970 (Biosphere Reserve since 1981).

"It's Worth a Trip - And So Is Staying": In this brochure the German biosphere reserves present themselves as holiday destinations.

A study of the relationship between day trips and overnight tourism revealed a dominance for day trips in the Bavarian Forest, Berchtesgaden and the Hamburg and Lower Saxony Wadden Sea BR. By contrast, in the Schleswig-Holstein Wadden Sea BR 15 million overnight visitors dominate in comparison to twelve million day trippers (PETERMANN, T., REVERMANN, C. 2002: 43-44).

Tourism and Nature Conservation – a Contradiction in Terms?

The relationship between tourism and nature conservation is characterised by mutual dependency. On the one hand, tourism in biosphere reserves in particular – and in national and nature parks – benefits from the attractiveness of nature and the countryside. On the other hand, overexploitation for the purposes of tourism can have a negative impact on this. And the negative consequences are many and complex: they range from massive traffic problems, e.g. those in the South East Rügen BR (since the summer of 1991 there have regularly been up to 15,000 vehicles per day), right up to the negative impact of tourism on the flora and fauna, mainly resulting from certain leisure activities. The increased consumption of resources (land, water, energy) and waste and sewage product also have a negative impact.

Correct behaviour during sport reduces damage to flora and fauna.

On the other hand, tourism also has positive effects for nature and the countryside: tourism can help to improve the image of and acceptance for pure protected areas for nature, such as the core areas of biosphere reserves, and for nature conservation measures.

Moreover, positive economic effects are expected for the region (e.g. increasing the value-added rate). In the Bavarian Forest BR it is assumed that the protected area has a clear positive effect on the regional tourism industry since 30 per cent of the visitors spend their holidays in the region due to the national park of the same name.

Against the background of the opportunities and risks outlined above, ecological protection goals are often in opposition to the tourism development goals. The forms of sustainable tourism developed by the World Tourism Organisation (WTO) can offer a solution here; these are very important for the buffer zones and transition areas in biosphere reserves.

The promotion of sustainable economic, social and cultural development in the transition area is an objective of the Seville Strategy of the Man and the Biosphere Programme (MAB) as well as the protective function of the biosphere reserves for the purposes of conserving biodiversity in the core areas (UNESCO 1996). The UNESCO MAB Programme together with its biosphere reserves thus offers the best conditions for tourism development geared towards these principles.

The MAB Programme sees biosphere reserves as an opportunity to interlink the efforts for sustainable tourism development by means of international cooperation. Biosphere reserves also play an active role in implementing case studies on the application the Guidelines for "Biological Diversity and Tourism" of the Convention on Biological Diversity (CBD).

Approaches for Solutions in Biosphere Reserves

Visitor management

The concentration of visitor numbers in sensitive areas in terms of times and sites leads to potential for conflict that can be solved by visitor management. This is based on an analysis of critical loads and risk potentials. A key aspect of this is visitor guidance by means of infrastructure planning (car parks, marked paths). Using a "honeypot strategy", which combines infrastructure and an attractive range of services, visitors can be successfully concentrated in particular areas and, at the same time, acceptance for access bans in the core areas and buffer zones can be increased (PETERMANN, T., REVERMANN, C. 2002). For example, a comprehensive visitor guidance concept has been developed in the Vessertal-Thuringian Forest BR, where the network of tourist paths has been reduced overall; in order to avoid conflicts of use a maximum of two types of tourism use are permitted per path (cf Chapter 4.6).

Experiencing nature and letting it live at the same time – an element of sustainable tourism

Cooperation instead of confrontation

The medium range mountain landscape of the Rhön attracts climbers, walkers, mountain bikers and aerial sports. On the one hand, this is associated with a great deal of tourist potential, but on the other hand, conflicts with nature conservation objectives and social conflicts between the various user groups cannot be ruled out.

Usage concepts that have been jointly drawn up are one solution: thus, on the initiative of the Bavarian administrative authority of the Rhön BR and the *Allgemeiner Deutscher Fahrrad-Club* e. V. (ADFC) [German cyclists' association], a mountain biking concept has been developed to identify routes that are attractive from a sporting point of view and are largely ecologically sound (BIOSPHÄRENRESERVAT RHÖN 2001).

Integrative concepts

Visitor guidance and community protection-usage concepts are suitable for making tourism development in biosphere reserves ecologically sound. However, development along the lines of sustainability requires the incorporation of social, cultural and economic requirements.

Since biosphere reserves are often part of a large tourist region, integration in local government and regional planning is necessary. The players involved, e.g. providers of tourist services, tourism organisations, BR administrations, planners and politicians and, not least, the local population, must be continuously involved in cooperative processes.

The foundation of practically all integrative processes is an analysis of strengths and weaknesses of the region concerned, from which the opportunities and risks of tourism development can be seen and priority fields for action can be derived. The implementation of the "European Charter for Sustainable Tourism in Protected Areas" has been tested in the Palatinate Forest-North Vosges Transboundary BR since summer 2003. The Charter was developed by EUROPARC for national parks, nature parks and biosphere reserves. Participating areas undertake to develop and implement a concept and a plan of action for sustainable tourism in the region involving all relevant players. Participation can be marketed as a quality feature with great effectiveness.

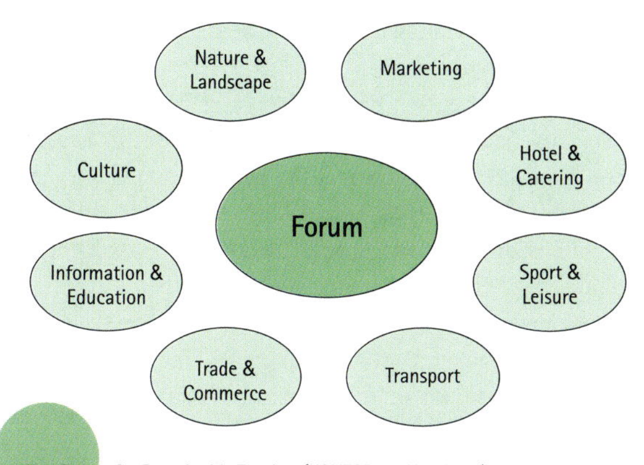

Fig.: Forum for Sustainable Tourism (KONTOR 21, Hamburg)

Prospects for the Future

Nature conservation is often considered to be a "brake" on economic development. The examples of sustainable tourism development in buffer zones and transition areas of biosphere reserves, however, show that this does not have to be the case. On the contrary: in biosphere reserves tourism is often an important economic foundation for the local population because biosphere reserves, as well as national and nature parks, convey a positive image and offer the possibility for a destination to present itself as unique for specific target groups in comparison to competitors.
Special importance is attached to the incorporation of tourism in concepts for sustainable regional development in rural areas with a poor infrastructure. This holds out the promise of various positive effects, e.g. more tax revenue, the creation of infrastructure, changes to payment flows and an increase in value-added rates in the region (PETERMANN, T., REVERMANN, C. 2002). When interlinking agriculture, skilled trades and tourism, an increasing number of biosphere reserves are turning to the marketing of local produce and the creation of their own regional brands, e.g. in the Schaalsee and Schorfheide-Chorin BR.
Biosphere reserves play a special role in the creation of jobs: under the motto "Biosphere Job Motor" the South East Rügen and the Schaalsee BR support entrepreneurs starting their own businesses and who want to implement the concept of sustainable economic development. Although in most cases only a small proportion of the company concepts relate directly to tourism, almost all of them benefit from the fact that Rügen and Schaalsee are popular holiday destinations (cf Chapter 4.3).

The tourist potentials of biosphere reserves have not yet been exploited. But the opportunities for the biosphere reserves lie in the creation of improved infrastructure resources and in providing specific, high-quality offers, e.g. tourism to experience nature or agricultural tourism, and the professional marketing of the same.
However, the future of tourism heavily depends on the quality of nature and the countryside. Polluted beaches and congested roads do not attract any tourists. The tourism industry has now recognised that holiday regions lose their attractiveness and thus the basis of their existence if they aim at quantity rather than quality.
But the joint development of objectives and concepts benefits not just tourism in the biosphere reserve and the models of sustainable development, but also the biosphere reserve itself: by increasing regional wealth creation, by giving greater importance to rural areas, by creating more acceptance for nature conservation among the local population and increasing awareness of nature and the environment among holiday-makers.

Literature

BIOSPHÄRENRESERVAT RHÖN (1996): Tourismus in der Rhön. Tourismus-Leitbild des Biosphärenreservats Rhön.
BIOSPHÄRENRESERVAT RHÖN (2001): 10 Jahre Biosphärenreservat Rhön – Zwischenbilanz einer Erfolgsgeschichte.
EUROPARC DEUTSCHLAND (2002): Biosphere Reserves in Germany: It's Worth a Trip – And So Is Staying.
F.U.R. (2000): Reiseanalyse 2000. Executive summary. Forschungsgemeinschaft Urlaub und Reisen e. V. Hamburg.
F.U.R. (2001): Pressetext "Sport im Urlaub – die Mischung macht's". Executive summary. Forschungsgemeinschaft Urlaub und Reisen e. V. Hamburg.
F.U.R. (2002): Reiseanalyse 2002. Executive summary. Forschungsgemeinschaft Urlaub und Reisen e. V., Hamburg.
F.U.R. (2003): Reiseanalyse 2003. Executive summary. Forschungsgemeinschaft Urlaub und Reisen e. V., Hamburg.
NEWSLETTER FAHRTZIEL NATUR. Ausgabe 19/3 17.07.2003.
PETERMANN, T., REVERMANN, C. (2002): TA-Projekt Tourismus in Großschutzgebieten – Wechselwirkungen und Kooperationsmöglichkeiten zwischen Naturschutz und regionalem Tourismus. Final Report., Büro für Technikfolgen-Abschätzung beim Deutschen Bundestag, Arbeitsbericht Nr. 77, März 2002.
VERBAND DEUTSCHER NATURPARKE (VDN), 2002: Nachhaltiger Tourismus in Naturparken – Ein Leitfaden für die Praxis.

3.3.5 *Environmental Management in Industry*

Frauke Druckrey

It is not only since the UN Conference on Environment and Development (UNCED) in Rio de Janeiro in 1992 that many companies in Germany have been attempting to operate on a sustainable basis. Environmental protection has been a leading priority for many years. On the one hand, there has been a number of serious problems, but on the other hand, economic development and job safeguarding have been less at risk (ICCA and UNEP 2002: p. 11 onwards).

The success achieved in recent years in controlling air pollution, sewage and waste problems, protection of resources and safety of manufacturing plants, transport and products is impressive. This can certainly be accredited to the relevant statutory regulations in place. The commitment of management to environmentally friendly industry and the corresponding introduction of management systems have played and indeed still do play a crucial role.

It goes without saying that environmental protection alone cannot be the main priority for companies. The need to be economically successful, which will ensure the future survival of companies and employment, is, of course, of equal importance. Many German companies are demonstrating that they have successfully mastered this "tightrope walk" challenge.

This publication by the International Council of Chemical Associations (ICCA) is an example of how industry has reacted to the challenges of the Rio Conference.

Their managers have to aim for the company's long-term safeguarding and it is their responsibility to foresee any necessary structural changes. Particularly for many small and medium-sized businesses, this is not easy. It makes initiatives, networks and working unions even more important which create common goals and make mutual assistance and sharing of experiences possible.

A good example is the "Responsible Care®" initiative in the chemicals industry. Worldwide, this industry sector has set itself guidelines which commit the participating companies to continually improving their health, safety and environmental performance and to reporting on this to the general public. It is clear that employees and other interested stakeholders concerned should be involved in the development of the implementation programmes. The possibility of expanding the initiative to include more social and economic aspects is increasingly being considered.

In its "Responsible Care®" Report (VCI 2002 and www.vci.de), the Chemicals Industries Association reports annually on the successes and challenges of "Responsible Care®" in Germany. There are similar reports at European and international levels (CEFIC 2002 and www.cefic.be; ICCA 2002 and www.icca-chem.org).

Many companies in the chemicals industry are small and medium-sized companies and so an important element of "Responsible Care®" is to provide mutual support and assistance in achieving the set objectives.

The contribution of this sector of industry to sustainable development is indeed exemplary and could be used as a model for other initiatives, such as the UNESCO Programme Man and the Biosphere (MAB). Companies in biosphere reserves also need long-term safeguarding, they need to operate in a way that protects the environment, include the stakeholders involved, accept personal responsibility and adopt good management systems.

To enable biosphere reserves to function economically on a sustainable, long-term basis, they must be managed with the same professionalism as commercial businesses, without neglecting their conservation and logistic research function.

Literature

CEFIC (2002): Annual Report CEFIC Responsible Care®.

ICCA (2002): Responsible Care® Status Report.

ICCA and UNEP (Eds.) (2002): ICCA Chemical Sector Report to UNEP.

VCI (2002): Responsible Care® Daten der chemischen Industrie zu Sicherheit, Gesundheit, Umweltschutz, Frankfurt.

3.4 Research and Monitoring in Biosphere Reserves

Doris Pokorny and Lenelis Kruse-Graumann

The Importance of Research in Biosphere Reserves

In the German UNESCO biosphere reserves research in particular has to answer the question as to how an ecologically, economically and socially viable use can be designed for Central European cultural landscapes. Research is supposed to help to initiate sustainable development of this kind in the biosphere reserves, accompany the regions on their way towards this development, answer questions and identify solutions to problems.

Research subject: the future of the cultural landscape

Subjects for Research in Biosphere Reserves

In principle, all research subjects that result from the guiding principle or framework concept concerned are relevant to a biosphere reserve. Research applies equally to the natural, social and economic sciences since ecological-nature conservation, economic and socio-cultural questions have to be dealt with. Multidisciplinary, interdisciplinary and transdisciplinary approaches are considered to be helpful.

In biosphere reserves research is mainly dedicated to drawing up:

- regional indicators, criteria and standards as well as approaches for solutions for sustainable development and land use,
- strategies for conservation and sustainable use of abiotic resources as well as biodiversity (including crop varieties and domesticated animal breeds),
- strategies for putting a value on landscapes with the aim of keeping and creating jobs, linking in ecological and economic potentials, issues relating to sustainable tourism and comparing competing demands for use of the landscape,
- fundamental principles for environmental education, communications and public relations.

The aim for every biosphere reserve should be a thematically balanced range of research topics.

The Preconditions for Research in Biosphere Reserves

The administrative authorities of the biosphere reserves usually initiate, organise and coordinate the research projects of others (research institutes, higher education institutions, specialist administrations – with support from local experts and volunteers). Together with the people in the region they identify relevant research topics. Harmonising the contents of projects is also important.

Research subject: criteria for environmentally friendly land use

Not all German biosphere reserves have their own research budget, which would be essential to place research work ideally in terms of location, contents and time. It helps to ensure the compatibility of data, avoid duplicated data collection and ensures that existing information is used in the best possible way. Research activities also need to be guided in a way that is compatible with nature and society.

Communication of the Results

Returning the research results to the region is an important task for the biosphere reserve administrations. This can be done through publication of research results in presentations, brochures, series of documents, press releases or on web pages.

Monitoring to Check the Success of Sustainable Development

Monitoring is a task for the biosphere reserves (cf Seville Strategy partial goal III.2., recommendation 4). Gradually, the aim is to introduce an integrated system of environmental monitoring (cf Chapter 5.3) that encompasses the entire ecosystem with its most important components and processes. It considers causes and effects, compiles the existing information, supplements it where necessary and jointly evaluates this information. This integrated environmental monitoring is a new approach and differs from standard practice that has so far had a more sectoral basis.
However, alongside economic monitoring, "social and societal monitoring" above all is essential for the comprehensive documentation of sustainable development. Suitable indicators have to be developed for this in order to describe the lives and livelihoods of people in biosphere reserves, explain changes and maybe also make predictions. A basic concept and first proposals for indicators were elaborated in 2001 at an international workshop in Rome on social monitoring in biosphere reserves (LASS, W., REUSSWIG, F. 2002). As a next step, model projects in individual biosphere reserves would be desirable to analyse the suitability of various indicators more precisely and to investigate their general significance. In addition to this, the methodological problems of collecting them need to be clarified. On the basis of existing considerations and proposals, research projects, which are linked with each other both nationally and internationally, are much needed. Thus UNESCO biosphere reserves will be able to justify their role as model regions for sustainable development.

What Research and Monitoring Can Do and What Not

Above all, if research is to move biosphere reserves forward and give impetus to their development, creative approaches need to be applied and specific proposals for solutions are to be found. Thus planning-oriented disciplines are of specific importance. Research results help to make discussions on the local level less emotional **and more rational**. However, they can in no way make or assume political decisions. Neither could research results be implemented by scientists. This is then up to the stakeholders (land users, local politicians, associations, etc.) in the region.

The function of the BR administrations is the interface between research and/or monitoring and their implementation and is thus essential for the transfer of knowledge in the region.

Outlook

Biosphere reserves are attractive as research areas because of their focus on applied research, their interdisciplinary data and information pool and their logistical support through administrative bodies. In future, they should be used as pilot areas to a greater extent than in the past.
Measures to be taken are:
- to attract the attention of research donors at European and Federal level by promoting biosphere reserves as both research topics and research partners,
- to initiate joint research and monitor projects in and between biosphere reserves,
- to set up a research framework for every biosphere reserve on the basis of the area's most relevant concepts and guiding principle as well as
- to establish partnerships between universities and biosphere reserves.

Social and economic sciences should be given a much higher standing than in the past since they are the source of most environmental problems.
The following issues are important:
- Methods of developing guiding principles as well as the elaboration and testing of suitable participation models, taking particular account of gender roles/gender issues.
- Developing and testing strategies for solving conflicts.
- Developing and testing strategies to increase the acceptance of environmental campaigns.
- Developing and testing methods that activate and promote "soft" resources (knowledge, experience, identity, tradition) in the region.

Georg Vogel

Individual decision-making processes - every day, thousands of times

Biosphärenreservat Rhön, Teilgebiet Bayern

(How) can we still arouse children's interest in their own environment?

- Developing and testing strategies to implement sustainable lifestyles and patterns of consumption as well as strategies to control individual and societal decision-making processes towards greater sustainability.
- Developing and testing communications strategies against the background of changing values ("the environment is not fashionable any more").
- Developing and testing ways and organisational structures appropriate for sustainable regional development.
- Developing a concept for "monitoring sustainability" in biosphere reserves.

Literature on the subject

GERMAN MAB NATIONAL COMMITTEE (Ed.) (1996): Criteria for Designation and Evaluation of UNESCO Biosphere Reserves in Germany, Bonn.
LASS W., REUSSWIG F. (Eds.) (2002): Social Monitoring: Meaning, Methods for an Integrated Management in Biosphere Reserves. Report of an International Workshop Rome, 2-3 September 2001, published in BRIM Series.
REGIERUNG VON UNTERFRANKEN BAYERISCHE VERWALTUNGSSTELLE BIOSPHÄRENRESERVAT RHÖN (Ed.) (2002): Forschung im Biosphärenreservat Rhön. Netzwerk für eine nachhaltige Entwicklung.
SCHÖNTHALER, K. et al. (2001): "Modellhafte Umsetzung und Konkretisierung der Konzeption für eine ökosystemare Umweltbeobachtung am Beispiel des länderübergreifenden Biosphärenreservates Rhön", R&D Project 109 02 076/01 on behalf of the Bavarian State Ministry for Land Development and Environmental Issues and the Federal Environmental Agency.
UNESCO (Ed.) (1996): Biosphere Reserves. The Seville Strategy and the Statutory Framework of the World Network, Paris.

3.5 *Planning for Biosphere Reserves*

Dieter Mayerl

If UNESCO biosphere reserves want to be convincing as model regions for sustainable development, environmental action and quality objectives must be drawn up for these areas. These objectives should be derived from the model of sustainable development focused on the future ("The Concept of Sustainability") (DEUTSCHER BUNDESTAG 1998) and should include ecological, economic and social aspects. In this respect, well-understood planning involving all of the people affected plays a key role. The administrations of the biosphere reserves have a responsibility for this. It is just as important, however, to include the objectives and measures for the individual biosphere reserves in the supraregional plans for *Land*, regional and local community planning (AGBR 1995). This could also include other possible regions, such as industrial urban areas.

Hausen - A village in the Bavarian part of the Rhön Biosphere Reserve that as become the first model municipality of the Biosphere Reserve.

Planning Necessary for a Sustainable Future

During a time when politics and society and others involved in a particular region are fundamentally sceptical and critical towards planning, the administrations of biosphere reserves need to work hard at convincing and informing people about the different plans. The people involved and the different social groups should be included in the planning for biosphere reserves right from the beginning. Above all, the convincing message should be that it is not just a question of planning for its own sake. Indeed, it must be made clear that the implementation of the planning guidelines from the UNESCO Programme Man and the Biosphere (MAB), the Seville Strategy (UNESCO 1996) and UNESCO recognition in conjunction with the National Criteria (GERMAN MAB NATIONAL COMMITTEE 1996) for a model region for sustainable development is absolutely necessary.
Criteria 17 to 20 of the National Criteria define clear standards:

Planning Criteria for the Biosphere Reserves Planning

(17) A coordinated framework concept must be prepared within three years after the biosphere reserve has been designated by UNESCO. The application must contain a commitment to provide the necessary funding.
(18) Maintenance and development plans should be prepared within five years, on the basis of the framework concept - at least for areas within the buffer zone and transition area that require particular protection or care.
(19) The biosphere reserve's aims and the framework concept should be integrated, at the earliest possible time, within Land and regional planning and within landscape and development planning.
(20) Aims for the biosphere reserve's protection, maintenance and development should be taken into account in updates of other technical planning.

(German MAB National Committee 1996: 8)
(See National Criteria in the Annex p.164)

Above all, an agreed framework concept in line with Criterion 17 holds great importance for the protection, maintenance and development of a biosphere reserve. As a result of regionalised models, the framework concept contains

- environmental quality objectives to fulfil the potential of the environment and
- environmental action objectives for the implementation of the measures required to achieve the described condition of the environment (UBA 2000).

They include the biotic and abiotic conditions as well as the anthropogenic effects resulting from the use of the land. The different zones of a biosphere reserve play a decisive role in setting the objectives. The description of the objectives should be practical and focused on implementation. The steps that need to be taken to achieve the environmental action objectives should lead to concrete measures for all fields of expertise and will be prioritised according to the need for action.

Open Round-Table Planning

These days, if planning is to be accepted and supported by those involved, it can no longer take place "behind closed doors". The principle of open round-table planning should be

Landscape planning that is isolated and detached from the interests of the citizens will soon come up against some limitations. (by Mueller/IMAGO 87)

applied here (MAYERL, D. 1996). Open planning means that all people involved are included in the planning initiatives from the very beginning and that they are convinced of the benefits of forward-looking planning. They can contribute their ideas and suggestions in working groups and play a part in the planning process in line with Agenda 21. Round-table working groups are an established means of taking different interests into account (BAYSTMLU 2002a). Ideas from other specialist planning, such as the use of the land by agriculture and forestry, should also be included here. Integrated solutions for conflict can be established in this way.
In this open planning process it is an important task of the administrative authorities or the representatives that they have appointed, to give a positive impression of the international and national guidelines.
This process needs to take place at the round table if the people living and working in the biosphere reserves are to adopt the objectives and measures and play a real part in the implementation. They should recognise during this process that their involvement can also be to their advantage. Their identification with the biosphere reserve will develop from their personal commitment and from the opportunities that they have to contribute their opinions and to get involved in the organisation.

Planning for the Current Network of Biosphere Reserves in Germany

To implement the objectives and tasks of the biosphere reserve and to help it gain wider acceptance, it is important to demonstrate their positive aspects at the different levels of planning. A distinction needs to be made between the following levels:

- **Local planning**, such as
  - the framework concept for the whole area and
  - the protection and development plans for parts of the area, e.g. the buffer zones or conservation areas.
- **Integration** into
  - the supra-regional *Land* and regional planning (development plans according to the *Land* statutory planning regulations),
  - the local landscape and construction management planning system (local authority plans according to the laws of the *Länder*).

Planning for biosphere reserves is binding at different levels. *Land*, regional and local authority planning (as a rule, landscape and construction management planning) is binding for all public organisations involved in planning. On the other hand, local planning for biosphere reserves generates

Example from the Bavarian Land Development Programme – continued in 2003

Objectives for biosphere reserves:
Part B: Objectives for the sustainable development of specialist fields that have an impact on significant areas of land
Objective 2.1.2: (...)
The requirements for UNESCO biosphere reserves shall be established in suitable landscapes through the protection of different regions
Objective 2.2.1: (...).
In areas that are recognised by UNESCO as biosphere reserves, the model effects of human activity shall be implemented through planning and measures relevant to the different zones.

(BAYSTMLU 2003, non-official translation)

Example from the Western Mecklenburg Regional Development Programme

Regional Development Programme: Objectives for the Schaalsee Biosphere Reserve Region

- *Protection of the extensive ecosystem compound of lakes and other wetland biotopes and adding linking elements to the agricultural landscape.*
- *Preservation of the extensive, largely intact and unspoilt areas of landscape in terms of protection of species and recreation.*

(REGIONALER PLANUNGSVERBAND WESTMECKLENBURG 1996)

At the beginning of plannings for biosphere reserves all groups involved are in the same boat. (by Mueller/IMAGO 87)

voluntary commitment for the specialist fields which are involved in the planning. The more extensive and detailed the scope of the involvement, the stronger the voluntary commitment will be. This can be applied to all parties involved in planning.
Local planning for the biosphere reserve stems from the international and national MAB guidelines and specialist requirements. Whereas drawing up the framework concept is mandatory, protection and development plans for parts of the biosphere reserve need to be compiled according to specialist requirements. This can apply to the buffer zone or to parts of it or to particular conservation areas in the biosphere reserve that are in need of care and attention.

Implementation of Local Planning for Biosphere Reserves

All practical and effective planning shows specific measures for its implementation. Using projects and measures for its implementation means that those involved understand and are well-informed about the planning.
The first measures should start to be implemented during the open planning and the discussions at the round table. Exemplary projects implemented early on promote acceptance of the biosphere reserve very effectively. The people living and working here are immediately able to experience the implementation – whether regional identity is strengthened and regional products are marketed or the protection of the landscape by landscape management associations is requested and respected appropriately. This is motivating and inspires confidence, as concrete sustainable results are visible early on. Joint action leads to success and encourages people to continue (BAYSTMLU 2002 b).
The implementation of local planning must be accompanied by compelling and consistent information from the administrations. This could be through public meetings, letters to the public, the relevant community newspapers, brochures and also the new media, such as the internet. The start of a model project should be communicated through effective public relations work and there should be continual information on its progress. Cooperation with the press is important here, to communicate the information events and report on the results.
The success of the implementation of the planning should be monitored and the efficiency of the respective projects for sustainable development should also be reviewed. This can result in the planning being continued and the measures and projects suggested in the planning being adapted or reappraised.
Effective planning with all those involved is an on-going task in biosphere reserves, which must respect current social, political and specialist requirements.

Example from the Rhön Biosphere Reserve

General regional framework for the Rhön Biosphere Reserve
The framework concept comprising three Länder for conservation, maintenance and development (BIOSPHÄRENRESERVAT RHÖN/GREBE, R. 1995) constitutes a successful example of extensive involvement by the population in the drawing up of plans in the biosphere reserve. The results from sectoral reports (agriculture and forestry; transport and industry; housing schemes) and supplementary Land planning reports have been integrated into the framework concept. This exemplary work was drawn up in the three Länder (Bavaria, Hesse, Thuringia) during a round-table discussion process over several years involving all the communities, specialist authorities, associations and societal groups concerned and many regional experts.
(ABGR 1995, GERMAN MAB NATIONAL COMMITTEE 1996).

The Rhön Biosphere Reserve Framework Concept

Literature

AGBR (Ständige Arbeitsgruppe der Biosphärenreservate in Deutschland) (1995): Biosphärenreservate in Deutschland. Leitlinien für Schutz, Pflege und Entwicklung, Berlin-Heidelberg, etc.

BayStMLU (Bayerisches Staatsministerium für Landesentwicklung und Umweltfragen) (2002 a): Blaue Box – Werkzeuge Landschaftsplan-Umsetzung, München.

BayStMLU (2002 b): Informationen zur Blauen Box – Landschaftsplanung effektiv umsetzen, München.

BayStMLU (2003): Landesentwicklungsprogramm Bayern 2003 (Fortschreibung), München.

Deutscher Bundestag (1998): Abschlussbericht der Enquete-Kommission "Schutz des Menschen und der Umwelt" des 13. Deutschen Bundestags. Konzept Nachhaltigkeit. Vom Leitbild zur Umsetzung. Drs. 13/11200, Bonn.

German MAB National Committee (Ed.) (1996): Criteria for Designation and Evaluation of UNESCO Biosphere Reserves in Germany, Bonn.

Mayerl, D. (1996): Landschaftsplanung am Runden Tisch – kooperativ planen, gemeinsam umsetzen. In: ANL (1996): Landschaftsplanung - Quo Vadis? Standortbestimmung und Perspektiven gemeindlicher Landschaftsplanung. Laufener Seminarbeiträge 6/96, Laufen.

Regionaler Planungsverband Westmecklenburg (Ed.) (1996): Regionales Raumordnungsprogramm Westmecklenburg, Schwerin.

UBA (Umweltbundesamt) (2000): Ziele für die Umweltqualität – Eine Bestandsaufnahme, Berlin.

UNESCO (Ed.) (1996): Biosphere Reserves. The Seville Strategy and the Statutory Framework of the World Network, Paris.

Realisation parallel to planning (by Mueller/IMAGO 87)

3.6 *Biosphere Reserves in Development Cooperation*

Monika Dittrich and Rolf-Peter Mack

Introduction

The supreme goal of development cooperation is to reduce poverty. To this end, natural resources should be preserved on a long-term basis so that everyone can enjoy a rewarding life today and in the future.

Development cooperation understands that nature can never be protected from people. It can only be preserved on an ongoing basis with and for the population. Clearly, a balance continually needs to be found between conservation and use. The UNESCO Man and the Biosphere Programme (MAB) combines human action with the preservation of nature. In this respect, as far as development cooperation is concerned, biosphere reserves represent an approach that is particularly worthy of support.

General Conditions in Developing Countries

Many of the worldwide hotspots in biodiversity are in tropical countries. In recent decades, large areas have been placed under protection. At a national level, some of them have been identified as national parks, in line with or based on Conservation Category II of the IUCN – The World Conservation Union, the world's biggest nature conservation organisation. At an international level, they have been recognised as biosphere reserves since the middle of the 1980s, often having the earlier national parks as their core area.

To date, the approaches to managing protected areas have been based on the following assumptions (cf Amend, S. et al. 2002: 28-30):

- The areas are state-owned.
- The areas are remote and uninhabited.
- The institutions can exercise their influence.
- The management has an ample number of well-educated people and sufficient financial means.
- There is political will for nature conservation.
- There are rigid laws in place for the protected areas.
- There is wide social acceptance for nature conservation.

However, in reality, these conditions do not always exist as such. Generally, the authorities and institutions responsible for conservation stand out because they have few staff, who are for the most part poorly educated, few financial resources and a weak administrative structure.

They often depend on international resources and their plans have to be implemented by external personnel, who are not familiar with the areas and their population. Due to the centralist structures, the park directors have only very limited decision-making power and few opportunities to implement measures.

The political and legal framework frequently appears confusing. Conditions for use and the external and internal boundaries for protected areas and biosphere reserves are continually in conflict with one another. Policies in other social sectors have a counter-productive effect on the work of the comparatively young and weak environment ministries. This applies to mining, forestry and agriculture, land laws and reforms and the rights of ethnic minorities.

Another characteristic feature is that the local population is poor and either overuses the small amount of available resources or the housing areas shift into the protected regions. The awareness of the need for conservation is not particularly high and the scope for action is primarily based on short-term use. Altogether, this leads to poor acceptance of conservation objectives.

Nevertheless, acceptance increases if conservation is embedded in regional development objectives, if alternative, sustainable forms of use are propagated and if authorised and the population is involved with the management, including decisions and usufruct (such as sharing the proceeds from tourism in conservation areas).

Although large areas with few inhabitants can still be found in developing countries, only seldom is there an inventory of nature and existing species. At the same time, the international pressure to develop available or presumed resources (wood, medicinal herbs etc.) is increasing.

It is often not until biosphere reserves are designated that governments in developing countries aggressively attempt to include protected areas in regional development objectives and to propagate alternatives for use that conserve nature.

Rivers in the Amazon area of Ecuador often represent the only means of communication. Transparency in planning and the setting up of biosphere reserves increases the acceptance amongst the riparian populations.

A long period of time is usually required before the concept of biosphere reserves is accepted in the regions. Nevertheless, in the medium term, these efforts are already increasing both the acceptance of conservation amongst the population and the political commitment of governments, including lower levels. Designation as a "UNESCO Biosphere Reserve" is more and more regarded as a sign of recognition for the region that can be worn with pride.

Approaches of Development Cooperation

By signing the Convention on Biological Diversity (CBD) in 1992, Germany committed itself to international cooperation in the implementation of measures for the protection and sustainable use of biodiversity. It also agreed to share the benefits resulting from biodiversity fairly and equally. In 1992 the German Federal Government declared its support for the objectives and principles of sustainable development, as set down in Agenda 21.

Ever since, the preservation of biodiversity has represented an integral component of the bilateral development cooperation of the Federal Republic. Biosphere reserves can indeed be regarded as specific areas for the implementation of the CBD. The discussions regarding the preservation of biodiversity within development cooperation go back as far as the 1980s. The first project supported by Germany that dealt with conservation in the narrowest sense and therefore had the preservation of biodiversity as its objective was the "Selous Conservation Project" in Tanzania, which began at the beginning of the 1980s (cf KASPAREK, M. et al. 2000: 18).

Since 1985 the Federal Ministry for Economic Cooperation and Development (BMZ) has supported around 360 projects which contribute to the protection and the sustainable use of biodiversity and to fair benefit sharing. The Federal Republic is one of the biggest donors worldwide. Today, the implementation of the CBD is linked to the Action Programme 2015, which plans to cut worldwide poverty by half by 2015 (BMZ 2002: 16-17).

Germany is pursuing the objective of improving the capabilities of national institutions, population groups and also poor individuals, so that they in turn can either bring about progress in their country or improve their own individual circumstances. An important principle is to include the local population. Participation, i.e. taking part in political, economic and social processes, is an integral part of the principle of worldwide conservation supported by development cooperation.

Photograph from the project work

The project entitled "Demarcation of Indian Conservation Areas in the Amazon" was established within the framework of the Pilot Programme for the Protection of the Brazilian Rain Forests. Demarcation of boundaries means finding an agreement between that which is represented on a map and the area that exists in reality.

With regard to development cooperation, the concept of biosphere reserves appears to be an approach that particularly deserves support. This is because it combines conservation and use on the one hand and it is aiming at preserving biological diversity with all its opportunities for future generations on the other hand. Rather than being seen as a disruptive factor, the local population is regarded as a fundamental, important player in the preservation of nature.

Worldwide Commitment to Biological Diversity

Since 1985, the Federal Ministry for Economic Cooperation and Development (BMZ) has supported 360 projects for the protection and preservation of biological diversity (biodiversity). 45 per cent of the projects were or still are in Africa, 32 per cent in Latin America, 18 per cent in Asia and the remaining projects are in the Middle East, the countries in transition or they are worldwide projects (BMZ 2002: 10-11). As far as individual countries are concerned, Brazil receives the greatest support, followed by Bolivia, Tanzania, Peru, Madagascar und then Ghana.

The German contribution to the Global Environmental Facility (GEF), set up for the implementation of international environmental conventions (the Conventions on Climate Change, Desertification and Biological Diversity), comes to a total of over € 600 million. Within the framework of GEF, Germany has provided approximately € 260 million to the conservation of biodiversity (BMZ 2002: 15).

Indian woman with her child. Cultural diversity goes hand in hand with the preservation of biological diversity.

Projects that have the objective of setting up or consolidating biosphere reserves and preserving biodiversity, adopt the following fundamental approaches:

- planning and (land) management in different types of areas,
- improving local economic development and sustainable management of resources,
- managing transnational biosphere reserves and conservation areas and setting up bio-corridors,
- economic assessment of environmental goods and services,
- institutional improvement of public and private/civil society organisations,
- implementing international conventions, such as the Convention on Biological Diversity (CBD),
- political advice at a national and sectoral level and
- coordinating the measures of international donors.

Literature

AMEND, S. et al. (2002): Planes de Manejo – Conceptos y Propuestas. Parques Nacionales y Conservación Ambiental, No. 10, Panama.

BUNDESMINISTERIUM FÜR WIRTSCHAFTLICHE ZUSAMMENARBEIT UND ENTWICKLUNG (BMZ), DEUTSCHE GESELLSCHAFT FÜR TECHNISCHE ZUSAMMENARBEIT (GTZ) GmbH (2002): Biodiversity in German Development Cooperation, Eschborn.

DEUTSCHE GESELLSCHAFT FÜR TECHNISCHE ZUSAMMENARBEIT (GTZ) GMBH, BUNDESAMT FÜR NATURSCHUTZ (BFN) & INTERNATIONALE NATURSCHUTZAKADEMIE INSEL VILM (Eds.) (2000): Naturschutz in Entwicklungsländern: Neue Ansätze für den Erhalt der biologischen Vielfalt, Heidelberg.

KASPAREK, M. et al. (2000): Naturschutz – eine Aufgabe der Entwicklungszusammenarbeit. In: DEUTSCHE GESELLSCHAFT FÜR TECHNISCHE ZUSAMMENARBEIT (GTZ) GMBH, BUNDESAMT FÜR NATURSCHUTZ (BFN) & INTERNATIONALE NATURSCHUTZAKADEMIE INSEL VILM (Eds.) (2000): Naturschutz in Entwicklungsländern: Neue Ansätze für den erhalt der biologischen Vielfalt, Heidelberg, p. 11 – 26.

3.7 The Further Development of the German System of Biosphere Reserves – Model Regions for Sustainable Development

Alfred Walter, Hans-Joachim Schreiber and Peter Wenzel

Being a model region for sustainable development is a very high ambition. Sustainable development is internationally defined as a form of development that meets the needs of the present without compromising the ability of future generations to meet their own needs. In biosphere reserves, this intergenerational agreement should be put into practice with the local people by protection and sustainable use of the biosphere, i.e. the inhabited world. In fulfilling this function, biosphere reserves are designed to offer examples and serve as models for sustainable development.

In future, every UNESCO biosphere reserve in the World Network of Biosphere Reserves must contribute to sustainable development and further disseminate their solutions – each from its own specific position, taking into account its economic, social and cultural context. The system of biosphere reserves must be directed towards this objective, with regard to its subject matter as well as its structure and organisation.

The 1996 Criteria for Designation and Evaluation of UNESCO Biosphere Reserves in Germany are of central importance to the system of biosphere reserves in Germany (National Criteria cf Annex, p. xxx). They specify the International Guidelines for the World Network of Biosphere Reserves adopted by UNESCO in 1995. The German MAB National Committee bases the recognition and periodic review of German biosphere reserves on these criteria. The criteria thus ensure implementation of the MAB Programme in Germany at a high level.

The formulation of clear requirements for the structure of biosphere reserves in the catalogue of criteria has proved its worth. The structural criteria concern the representative nature of the ecosystems, the area size, the specification of the zones, the provision of legal safeguards, the establishment of a working administration and organisation and development of a framework concept.

However, the "functional criteria" laid down in the national catalogue of criteria, still need to be further developed – especially with regard to the greater involvement of economic and social aspects. In particular, the requirements to meet the development functions need to be further established along the lines of the new concepts described above. The subjects of education for sustainable development, participation and preserving the cultural and traditional identity of the region should form a new focus in the national criteria and, thus, in the future work of the German biosphere reserves.

In Germany there are currently 14 UNESCO biosphere reserves. They represent various ecosystems and landscapes. With regard to the regional situation they are facing very different challenges. Not only does the MAB Programme intend to create model projects that offer examples to other biosphere reserves and the areas surrounding the biosphere reserves. It is also working on the development of new solutions for other regions and other countries. Therefore, it is vital that not only the various natural areas be represented, but also the varying social and economic conditions. As a consequence, a completion of the German network of biosphere reserves would certainly be a biosphere reserve in an urban area, a biosphere reserve in a post-mining landscape and a biosphere reserve with raw materials extraction.

Impressions of the Rhön Biosphere Reserve

Under the UNESCO guidelines the MAB National Committees serve as focal points for the national concerns of the international programme. In the next few years, the main task of the German MAB National Committee, as UNESCO's executing body, will be to ensure a high level of implementation for the MAB Programme in recognising and periodically reviewing biosphere reserves. The National Committee plays a key role in developing new concepts.

In Germany, which is a federal country, the *Länder* are responsible for executing the implementation of the MAB Programme. The periodic review of the biosphere reserves in Germany has shown that efforts in biosphere reserves were focused on the conservation of nature and the countryside in the past and that exemplary work has been done here.

The realisation of the new concepts described above needs the adoption of additional tasks and a shifting of tasks within the administrations. For this, it appears necessary to direct the resources of the biosphere reserve administrations to the new requirements and tasks, both in terms of quality and quantity. In the administrations, thorough understanding of biology is as well required as business management, sociological and educational knowledge.

The heads of the German biosphere reserves constantly provide each other with new information. This lively exchange of experience has proved very valuable: it improves efficiency in implementing the MAB Programme and prevents duplication of effort, leads to coordination in dealing with interdisciplinary issues and creates important, technical foundations for the work in the *Länder* and in the MAB National Committee.

Germany also has to make its contribution to the further development of the international programme. Above all, this is a task involving the Federal Foreign Office and the MAB National Committee.

If the World Network of Biosphere Reserves is to meet the demands of being a network of model regions for sustainable development, it seems to be essential for UNESCO also to focus on the subjects of quality economies, education for sustainable development and socio-economic monitoring.

Bilateral cooperation plays an important role in the MAB Programme. Almost all biosphere reserves in Germany have bilateral contacts with areas in other countries. There is an intensive exchange of experience and mutual support.

The bilateral cooperation between the German and the Chinese National Committees also leads to a regular exchange of experience between biosphere reserves (cf Chapter 4.13). In future, the cooperation will be focused on sustainable regional development and environmental education.

Development cooperation will play an important role in establishing and preserving the UNESCO World Network of Biosphere Reserves. For example, German development projects can be found in the biosphere reserves of Arganie in Morocco, Bosawas in Nicaragua and Issyk-Kul in Kyrgyzstan. It would be desirable for the concept of biosphere reserves to be used as a standard instrument of German development cooperation in future (cf Chapter 3.6).

Examples from Practice

4.

4.1 *From the Rhön Lamb to the Rhön Apple Initiative: Marketing Local Produce*

Biosphere Reserve Rhön

Michael Geier

General Conditions for Marketing Local Produce in the Rhön Biosphere Reserve

Starting point in 1991

The Rhön Biosphere Reserve with its three parts in three different Federal *Länder* is not an entity that the population of the Rhön has actively worked for. Rather, the application for entering the UNESCO Man and the Biosphere Programme (MAB) and the geographic demarcation go back to an initiative at state level by the former German Democratic Republic for the Thuringian part and two nature conservation associations in the Hessian and Bavarian part of the future biosphere reserve. Until the Rhön was recognised as a biosphere reserve by UNESCO in 1991, the population of the Rhön was neither involved, nor informed in any detail about the process.

At the time the Rhön was recognised as a biosphere reserve, food retailers had already started to move away from the villages. An increasing number of villages were losing their grocers, bakers and butchers. Suppliers of goods of daily needs were increasingly moving to larger village centres or to the nearer district towns. At this time the marketing of Rhön produce in and outside the Rhön was but of marginal significance.

General market conditions

The purchasing power of the Rhön population always lay well below the national average. The supply of needs that were beyond the essentials was and still is to a large extent dependent on the price.

For a long time trade relations outside the Rhön were largely impeded by the considerable distance from supra-regional urban centres, such as the Rhine-Main area or the town of Würzburg.

Material and non-material support

In 1991 fresh impetus came from the implementation of the European Union's (EU) Structural Support Programmes (under Objective 5b) and the European Community's initiative LEADER. The implementation of these support programmes, fundamentally new in terms of their approach and make-up, required new organisational structures. In Hesse, the *Verein Natur- und Lebensraum Rhön* (Rhön Nature and Living Space Association) was set up as a so-called local action group under LEADER. In Bavaria, interest groups were established on the initiative of the "Rural Development Group in the 5b area", a working unit within the Lower Franconia government, implementing the "bottom up" approach as requested by the EU. With the introduction of LEADER II, both local action groups on the Thuringian side have mainly focused their support on the Rhön in order to overcome the negative impact of the former German-German border.

Apple trees in the Rhön Biosphere Reserve

Establishing and Strengthening the Marketing of Local Produce

Impetus by non-state organisations

One initial impetus that is very important until today came in 1984, i.e. long before the recognition of the Biosphere Reserve by UNESCO: The Bavarian *Bund Naturschutz BUND* [a non-governmental nature conservation association] launched the Rhön Lamb Project at that time, which since has become a successful self-supporting project.

The Verein *Natur- und Lebensraum Rhön* (Rhön Nature and Living Space Association), founded in 1991 to support the Rhön Biosphere Reserve, has tried from the outset to drive forward the regional and supra-regional marketing of Rhön produce. The most recent initiative to this effect was initiated by the *Arbeitsgemeinschaft Rhön* (Rhön Cooperation Group), ARGE Rhön for short, with the development and introduction of the umbrella Rhön brand.

The role of the governmental and the Biosphere Reserve administrations

The possiblities for the Biosphere Reserve administrations to actively support marketing strategies were, and still are, limited. None of the three administrative authorities are officially in charge, neither in an advisory nor a financial capacity. The administrations had to limit their work to personal and organisational support of the people actively involved, the non-governmental organisations and the farmers.

Current Situation of Local Marketing

Product range

In the last twelve years since the recognition of the Rhön as a biosphere reserve – and the start of the EU Structural Support Programmes – the marketing of local produce within and outside the Rhön has enjoyed a remarkable improvement. Genuine product innovations have been added to the existing range. Specific campaigns, taken particularly by the *Verein Natur- und Lebensraum Rhön*, have resulted in the creation of a range of Rhön produce.

Main Rhön Produce:

Rhön lamb · Rhön apples · Rhön outdoor-reared beef · Rhön brown trout · Biosphere beef · Rhön caraway seed bread · Rhön quality honey · Rhön goat produce · Rhön organic dairy produce · Rhön organic beer · Rhön wood products

Participants and networks

Local or regional groups of producers, such as the Rhön farms or the Rhön lamb cooperatives, were formed early on. Complex networks developed, such as the Rhön Apple Initiative or the Rhön Wood Processors bringing together those involved at all working and marketing levels, starting from the producers of the raw material.

Since 1998 the *Verein Natur- und Lebensraum Rhön* has developed a working partnership system for businesses prepared to adhere to particular quality standards in production and to cooperate with each other.

New Rhön products: "Bionade", "Öko Bier+Apfel"(Organic beer+apple)

Increment value and jobs

As a considerable success can be regarded if a close and stable cooperation of shepherds with a local butcher results in economic benefit for both sides, as in the case of the Rhön shepherds Josef Kolb in Ginolfs or Dietmar Weckbach in Wüstensachsen. At a much more significant level, as far as turnover is concerned, this can be shown through the cooperation between the Rhöngold dairy and the Agrarhöfe-farms in Kaltensundheim.

In the best case scenario, the business enjoys so much success that additional staff will be taken on and considerable investments can be made. Examples of this are the Rhön exhibition apple press in Seiferts, Christof Genssler's organic farm bakery in Poppenhausen, the carpenters' workshop ('Hand-Wood-Heart') in Gersfeld, the ÖLV Rhön farms or the Pius Korb business in Unterweissenbrunn.

Furniture made from apple-tree wood

Potentials for the Future

Expansion of the product range

In the author's opinion, the considerable potential to market local produce in the Rhön has yet to be fully exploited. As far as food is concerned, this applies to poultry, ewe's milk, vegetables and spices.

Moreover, the production of fuel from oil seed rape represents an important and exploitable area for growth in the Rhön.

Sales within the region

To date, it has not been possible to introduce local produce into the regional large-scale catering sector to any large extent. However, considering the numerous large treatment and rehabilitation establishments, i.e. health clinics, located in both of the Bavarian Rhön administrative districts, it is clear that the sales potential is largely unexploited.

Spatial expansion

The more the present positive development continues, the more the export to urban centres will become necessary. In addition to the direct marketing initiative undertaken by a market stall in Frankfurt, day trippers and long-stay holidaymakers in the Rhön Biosphere Reserve represent another important potential of customers.

4.2 The Wilderness Camp on Falkenstein

Bavarian Forest Biosphere Reserve

Susanne Gietl

A Camp for Everyone

The wilderness camp in the Bavarian Forest Biosphere Reserve lies at the foot of Falkenstein. At 1,315 metres above sea level, Falkenstein is one of the most prominent mountains in the Biosphere Reserve.

Created to promote environmental education, the wilderness camp has been designed and built for children and young people above all. It was already quite unique at its conception. The first designs did not originate on the desk of an individual planner; they were the result of a creative workshop held precisely at the spot where the finished camp would later be. The very different ideas of students and teachers, architects and construction workers, biologists and educationalists, neighbours and strangers, both young and old, were incorporated in the planning process. This resulted in the designs for the different lodges (the meadow bed, the cave in the ground, the tree house, the water hut, the forest tent and the light house) to be used as living and sleeping quarters.

The camp is at the foot of magnificent Falkenstein

The State Structural Engineering Office in Passau transformed these different ideas into a coherent architectural concept. The goal of the concept was to unify nature and architecture. The individual buildings of the wilderness camp are sited in a forest clearing, on the edge of the forest, in the forest or next to a stream, according to their particular characteristics. The whole structure has been integrated into the natural surroundings of the area. Many forms of the construction include profiles and curves and also deficiencies of the actual topography. The individual buildings are as follows:

The central building

The central building fits into the contour of the terrain and is used for social functions and hospitality. The building was built with a modular, wooden post construction. The green roof rounds off the harmonious incorporation in nature.

The central building

The meadow bed

The meadow bed is built on the meadow and is perfectly camouflaged by a green roof. From inside the building, six large windows permit a good view of the grass, flowers and animals in the immediate vicinity.

The earth cave

The shape of the earth cave resembles a hemisphere. The inside of the cave can be reached via a tunnel-like entrance. Even in severe weather conditions the internal climate here remains stable. The lighting conditions and the earthy smell of the dwelling give visitors the feeling of being surrounded by earth.

The cave in construction

The tree house

People who live at the top of a tree feel the air around them and sway gently in the wind, like in a bird's nest. The living and sleeping levels are approximately eleven metres high. Access to the entrance of the hut is via a wooden ladder.

The water house

The water house stands on oak posts directly over the Geiselbach. The babbling, gurgling and sparkling of the stream is characteristic of the unique atmosphere surrounding the house.

The water house in the wilderness

The forest tent

This tent-like hut is shaped like the silhouette of a treetop. In the shade of the forest the day light is usually refracted as it streams through the large window niches of the hut. The lack of light inside the forest is tangible.

The light house

Light energy penetrates the glass and, depending on its composition, produces different colour effects.
The sleeping areas are arranged according to the four points of the compass and have been designed in different shades.
This hut will be completed in 2003.

The Natural Resources of the Region: Ecology under Construction

The wilderness camp on Falkenstein sees itself as an exemplary ecological and environmentally friendly project. The planners, with their choice of building materials, are following only the most outstanding methods of construction that promote ecology. For example, they are almost exclusively using regional building materials such as wood, granite, clay or glass. An important aspect of staying at the camp is to organise and experience daily life while being totally aware of what is going on round about.
Clear cycles of activity, such as supply and waste management using a compost and recycling system, the reed sewage treatment plant and the solar and photovoltaic system, raise

important questions and issues which can be taken up and worked on by the visitors and employees of the camp together. Depending on the season, catering mainly consists of regional produce from organic farms.
A regional collaborator, the "Der Pausnhof" organic hotel, is responsible for this service. It runs its own organic farm in line with the guidelines set by the "Biokreis e. V." association.

Meeting in the Camp

The wilderness camp on Falkenstein is also seen as a site for international understanding. Due to the location of the biosphere reserve, it offers a framework for international, but above all German-Czech events. For this to happen, different partners need to work very closely together, such as the regional youth group and local youth work in the Regen district or the Sumava National Park, the Czech neighbour of the Bavarian Forest Biosphere Reserve.
The wilderness camp is also a place of learning for adults and families. They use the camp's facilities for further education, workshops, excursions, discussion forums and other events.
On 31 May 2002 the Bavarian Forestry Minister, Josef Miller, opened the wilderness camp on Falkenstein. Numerous guests of honour and interested local citizens came to celebrate the completion of this long-awaited new establishment for environmental education.
On the day of the camp's opening, the main participants – the children in the 4th year of the local primary school – made sure that the camp was full of infectious enthusiasm and that a true "regional celebration" took place. They inspired others with the musical that had been composed especially for this occasion.

Learning about Nature

Environmental education is a core task of the Bavarian Forest Biosphere Reserve. As it appeals especially to older children and teenagers, the wilderness camp on Falkenstein is an important addition to the educational establishments that already exist, such as the information centre, the forest youth hostel or the forest play ground. All aspects of nature and an emerging forest wilderness can be thoroughly and closely experienced by young people.
In addition, the wilderness camp provides young people with the opportunity to get to know and be inspired about a world that is almost unknown to them today. The emerging forest wilderness can help them to experience the pleasures of seeing and observing, hearing and listening and touching and feeling.

Internet

www.wildniscamp.de

4.3 The "Biosphere Job Motor" – a Start-Up Initiative

South-East Rügen Biosphere Reserve

Michael Weigelt

In the South-East Rügen Biosphere Reserve the high unemployment in the region is used as an opportunity to set up new companies in order to establish a network of partners for the Biosphere Reserve administration in many different sectors. As well as the sponsors of education, the organisation involves an advisory council, comprising ministries, public agencies, chambers, associations and other institutions which can simplify administration and solve problems at the "round table" in a way that is otherwise impossible.
To date, 89 participants have received mentoring in five courses. The success rate is over 70 per cent.
A new part of the "Biosphere Job Motor" is the "Junior Biosphere Job Motor", giving young people the opportunity to set up their own companies in their home region. The Biosphere Reserve is seen as an opportunity; environmental education is used with a very practical approach here. The project has been running since 2002 at Sellin intermediate secondary school.
The "Biosphere Marketplace" has been developed as a common market for products from sustainable farming in biosphere reserves and other large protected areas, both nationally and internationally. It is the first approach for an international network on an economic basis. This will lead to new opportunities to create new jobs in the "Biosphere Job Motor". The most important foreign partner is currently the central national park administration of Columbia ("Café Biosphere"). Other partnerships are currently being set up.

At the 6th Rügen Wood Fair (2002): Agriculture Minister (Mecklenburg-Vorpommern) Backhaus and Ms Kassner, chief executive of the district, at the stand of the Columbian national park administration, which is being visited by a travelling craftswoman in the Posewald project. An almost symbolic image for the "Biosphere Job Motor" network

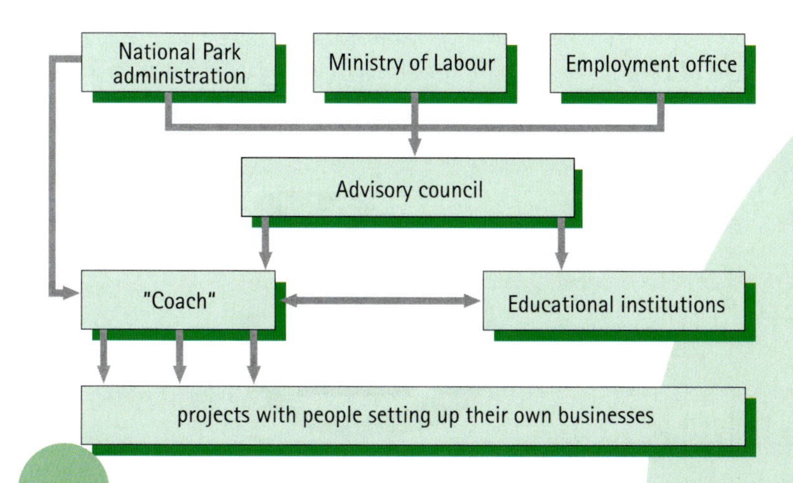

Fig.: The organisational structure of the "Biosphere Job Motor" in the South-East Rügen Biosphere Reserve (Source: Michael Weigelt)

The "Posewald Project School", an international meeting centre for young people in the Biosphere Reserve in the listed manor house with an "ecological industrial estate" (identical to the start-up centre of the "Biosphere Job Motor") in the surrounding area, is designed along the lines of the Danish production schools in order to offer young people opportunities to decide on a future career.

The concept has already been realised in the restoration of the site, in the form of educational schemes that also implement the aspect of sustainability when renovating listed buildings, e.g. in partnership with the travelling craftsmen.

Since 1997, the annual "Rügen Wood Fair" has illustrated how diversely the regenerative raw material of wood can be used and has explained the forest ecosystem and forestry. Widened into the "Wood and Regional Fair" and in conjunction with offers for the entire family, it is the scene for "Biosphere Marketplace" and "Biosphere Job Motor" as well as a "shop window" for the "Rügen Model Region".

The pun "biosfestival" is an umbrella term for cultural events in the Biosphere Reserve. To date, these have included the "blue boat" International Youth Jazz Festival and the Putbus Festival. The ideas come from the "Biosphere Job Motor", the link to the "Biosphere Marketplace" is mandatory.

The "Biosphere Ticket" is a modular system in which various tourist offers can be combined according to the individual wishes of the visitors. Services from the "Biosphere Job Motor" are interlinked with offers from local public transport. Current elements include "Thomas Trojan Sea Kayak Trips" and the "Rad & Heu" [Wheel & Hay] working group with seven new entrepreneurs from the "Biosphere Job Motor" to date. Added to this is the "Sea Eagle Touring" company currently being founded by schoolchildren from the "Junior Biosphere Job Motor".

The "Rügen House" in Zirkow, the home of the "Biosphere Job Motor", has already become an innovative start-up centre for the Biosphere Reserve and the Rügen district. Furthermore, it is being developed into the "House of the Biosphere" (regional cuisine, information, "Biosphere Marketplace", events, offers for children, local museum).

The Rügen House in Zirkow: home of the "Biosphere Job Motor"

In their entirety, all of these projects should be seen as a strategy for sustainable development, as the Agenda 21 of the Biosphere Reserve and its surrounding area.

4.4 *The Regional Brand as a Working Instrument for Sustainable Regional Development*

Schorfheide-Chorin Biosphere Reserve

Eberhard Henne

The Schorfheide-Chorin Project

One result of the Schorfheide-Chorin Project, entitled "Nature Conservation in the Agricultural Cultural Landscape using the example of the Schorfheide-Chorin Biosphere Reserve" (cf Chapter 5.6), funded by the Federal Ministry for Education and Research (BMBF) and *Deutsche Bundesstiftung Umwelt* (DBU) [a nationwide environmental foundation] was the development of model projects, one of which will be described in more detail below.

Goals of the "Regional Brand" Model Project

The establishment of a regional symbol of origin is based on much preliminary work and similar trials in the Biosphere Reserve. The initiators wanted to stimulate greater demand for products and services from the region that have been produced in an environmentally sound way by using a regional symbol of origin, the regional brand of the Schorfheide-Chorin Biosphere Reserve. The attractiveness of the Biosphere Reserve and the self-confidence of its inhabitants were to be strengthened, new activities encouraged in the sustainable development of the area.

The establishment of production and processing chains in conjunction with direct and regional marketing aims at increasing regional wealth creation.

Logo of the Regional Brand

Due to the high degree of familiarity of the Schorfheide-Chorin BR, it seemed to make sense to use the existing logo of the Biosphere Reserve. It has been registered as a picture mark in the register of the German Patent Office.

The use of the logo of the Schorfheide-Chorin BR was also to clearly illustrate the identification of the companies involved with the goals of the Biosphere Reserve.

The *Förderverein Kulturlandschaft Uckermark e. V.* (Cultural Landscape Uckermark support association) acts as the awarding body for the new regional brand.

Logo of the Schorfheide-Chorin Biosphere Reserve

The regional brand of the Schorfheide-Chorin Biosphere Reserve

Awarding and Criteria

It was necessary to draw up a brand statute that would govern the responsibilities and the procedure for awarding the logo and, at the same time, lay down criteria for the use of the logo. Under the statute, the Board of the *Kulturlandschaft Uckermark e. V.* association will appoint a permanent "Regional Brand" technical advisory committee, comprising two representatives from the Biosphere Reserve and two from the association.

The user of the regional brand must have a company based in the Biosphere Reserve or the majority of its products or produce must be produced on land within it.

Quality criteria have also been developed for the individual sectors of the regional brand so that the brand does not merely emphasise the origin of the products and services, but takes special account of issues of environmental protection and animal welfare as well as consumer safety and demand measures for a socially compatible way of doing business.

The use of the regional brand and its further development should be an incentive for the businesses to apply extensive methods of use that conform with nature conservation.
The criteria for the hotel and restaurant trade also contain conditions relevant to the environment.
The criteria are constantly updated and further developed in line with new findings in nature and environmental conservation.

Monitoring

Compliance with the quality requirements and the regional origin of the products are test criteria that are monitored by various recognised and independent institutes. In the restaurant, hotel, crafts, fishing and bee-keeping sectors, monitoring is only by the "Regional Brand" technical advisory committee of the *Kulturlandschaft Uckermark e. V.* association.
In April and May 1998 the procedure for the use of the regional brand was presented and discussed in the individual sectors.
In the summer of 1998 the independent testing institutes and the members of the technical advisory committee inspected 32 applicants. After the inspections, the technical advisory committee made a positive decision on 26 applications and the logos could start to be awarded.
Every applicant concluded a contract on permission to use the logo with the *Kulturlandschaft Uckermark e. V.* association. In it, the obligations for the user and the association are regulated so that the symbol of origin cannot be abused.
The association sends the user of the logo enamel signs, disk templates and product stickers with the regional brand.

Users

The following sectors were represented among the first 26 regional brand users:

Agriculture	7
including organic farms	*3*
Horticulture	2
Food processing	6
Skilled crafts	1
Fishing	1
Bee-keeping	5
Restaurant/hotel	4

The first logos for the regional brand were awarded on 27 August 1998.
After almost five years of application, almost 60 companies now use the regional brand of the Schorfheide-Chorin BR.

Outlook

The introduction of a regional symbol of origin and establishing regional economic cycles and a nationwide marketing system take a long time and are complex. Intensive and professional public relations work remains essential for a regional brand. The most important result of the regional brand project, however, is that the region develops greater self-confidence and starts new activities. In this connection, the protection of habitats and species are an integral element of the overall process in this cultural landscape.

Literature

FLADE, M., PLACHTER, H., HENNE, E. & K. ANDERS (2003): Naturschutz in der Agrarlandschaft, Ergebnisse des Schorfheide-Chorin-Projektes, Quelle Meyer.

4.5 The Framework Concept as Regional Agenda 21

Schaalsee Biosphere Reserve

Klaus Jarmatz

Introduction

From the outset, the UNESCO Man and the Biosphere Programme (MAB) of 1970 had the responsibility of developing and testing models for smooth cooperation between humans and nature. This goal was defined more precisely in the Seville Strategy (UNESCO 1996), drawn up in 1994. This recommends devising and pursuing concrete, regionally focused strategies in framework concepts, in order to implement the range of responsibility at a regional and national level. At the same time, there is the challenge of making positive use of the radiation effect of the model regions within biosphere reserves.

The Starting Point

UNESCO's recognition of the Schaalsee Biosphere Reserve in January 2000 resulted in the task of drawing up an agreed framework concept by 2003. Within this, there should be a balanced representation of the conservation of biological diversity and sustainable socio-economic development.

The large conservation area, established ten years before, could benefit from experiences in planning and cooperation activities. Since 1992, UNESCO's conservation function has been implemented and the area has been divided into various zones with the help of a specialist conservation plan. This was supplemented by ecological and socio-economic reports etc. and cross-border participation with the Lauenburg-Schaalsee landscape in the Federal Support Programme "Areas of National Importance". The framework concept could not have been drawn up before the deadline without clearly defined requirements for conservation.

Concurrent with the region's involvement in the MAB Programme, since 1993 considerable efforts have been made to promote both sustainable regional development and work on environmental education and information in the Schaalsee region. Even more significant is the fact that this has resulted in the formation of the first working groups to enhance regional communication.

The exhibition and the large media room in the "PAHLHUUS" provide what is necessary for regional and national events, taking care of visitors and providing information. In the first five years since its opening in 1998 the exhibition has had approximately 325,000 visitors.

AfBR Archive

The Organisational Structure

The framework concept for the Schaalsee Biosphere Reserve should not be directed internally. It should be devised using a cooperative and consensus-orientated approach and under fixed general conditions. To this end, the committee, a regional advisory council consisting of district administration heads, the Ministry for the Environment and Agriculture of Mecklenburg-Vorpommern, local representatives, associations and organisations, decided to set up a Regional Agenda 21.

A special feature of this is that projects that have been tested by experts should be implemented in parallel to the planning process. This has been achieved with the joint support of and financing by the two administrative districts, the community and the Biosphere Reserve administration. The running of the project and the management of the support funds (*from the Land* and the EU) were assigned to the association responsible for the biosphere reserve.

The Participation Model

In order to get as many citizens, interest groups, participants and others involved as possible, the project initiators developed their own model for participation. A steering group consisting of representatives from the supporting community and association, the leaders of the working groups and an environmental and regional planning representative managed the project in terms of making decisions about projects, internal focus etc.

Models and projects have been developed and agreed upon in open working groups, which have been supported by intensive public relations work. They represent the main focus of activity for the internal structure and future development of the framework concept. Issues such as tourism, housing estate development, investment in land, young people, welfare and energy have been dealt with.

A regional dialogue forum has been set up to guarantee discussion that covers interdisciplinary *Land* and working group issues. It consists of spokespersons from the working group and representatives from politics, management and various interest groups. It quickly gained acceptance from all participants.
To assist the project from an organisational point of view, a project office was set up with additional staff in the BR administration. The minutes of all committee meetings and the progress made in planning can be seen in a specially-designed internet presentation.

Since 2001, the working groups (in the above case, the Tourism Working Group) have been focusing on their goals and working very hard. In the Tourism Working Group alone, more than 50 participants have taken the opportunity to contribute their ideas for developing the region.

Results, Problems, Recommendations

As a result of the process, the agreed framework concept is divided into three volumes: "Analysis of Existing Developments", "Models and Goals" and "Action Plan/Project Overview". The plan's structure complies with the recommendations made by EUROPARC Deutschland (EUROPARC Deutschland 2000).
The project needs to survive into the future and so its principles and structures will continue to exist. The goals and models will be the benchmark for all future projects. Some project ideas have in fact already been implemented.
Important results from the process include balancing essential requirements and different poles of interest, a stronger identity, the development of an environment that encourages debate and structures for communication and cooperation. One successful outcome of this process was qualifying as a LEADER-plus-Region.

Such a complex process cannot be without problems. These, however, were solved with the help of a goal-orientated presentation and intensive conviction. All in all, the framework concept, drawn up democratically with the views of the citizens in mind, is seen as exemplary.
The process successfully achieved a balance between the different poles of opinion, strengthened regional identity and established networks and collaboration. The *Land* of Mecklenburg-Vorpommern recognised this achievement by awarding the first prize in the 2002 Environmental Competition.

Literature

EUROPARC Deutschland (2000): Leitfaden zur Erarbeitung von Nationalparkplänen, Berlin.
UNESCO (Ed.) (1996): Biosphere Reserves. The Seville Strategy and the Statutory Framework of the World Network, Paris.

4.6 Tourism with Nature – Nature Conservation with People: Visitor Guidance in a Biosphere Reserve

Vessertal-Thuringian Forest Biosphere Reserve

Johannes Treß and Elke Hellmuth

Introduction – Orientation for Joint Work

The Vessertal-Thuringian Forest Biosphere Reserve is located in the Central Thuringian Forest and is an important component of a tourist region with a tradition going back well over a hundred years. Accounting for around 50 per cent of the overnight stays in Thuringia, this region – the Thuringian Forest – is the most important tourist region in the *Land*. Even in the former German Democratic Republic the region was one of the most important areas for tourism. However, after 1989 there was a dramatic collapse in the number of visitors and overnight stays because many of the regular guests used the new freedom to explore holiday regions previously unattainable to them.

In the 1990s, towns, municipalities and the tourism industry made great efforts to revive the Thuringian Forest as a tourism destination. They undertook many activities to improve the infrastructure, offers and marketing of the region. In many cases, these activities were not very well coordinated and did not pay enough attention to other concerns, e.g. those of the forestry industry and of nature conservation. The following question arose: How can the concerns of tourism be brought into harmony with those of nature conservation in a well-developed and established tourist region? Against this background, the "Visitor Guidance" project was launched in 1999 to promote sustainable tourism development.
All players whose interests and concerns are affected by tourism were involved in an open dialogue and working process from the outset in order to ensure the sustainability of the project. In this way those involved drew up and realised the goals and measures of visitor guidance in a process of partnership.

Jointly Developing the Goals – Goals for an Entire Region

First of all, the goals of visitor guidance in the Vessertal-Thuringian Forest Biosphere Reserve had to be defined. An undergraduate thesis dealt with this first part of the project (KLEINE-HERZBRUCH, N. 2000). During this phase, two consultations were held with representatives from the region at which ideas were discussed and suggestions recorded (cf Fig.).

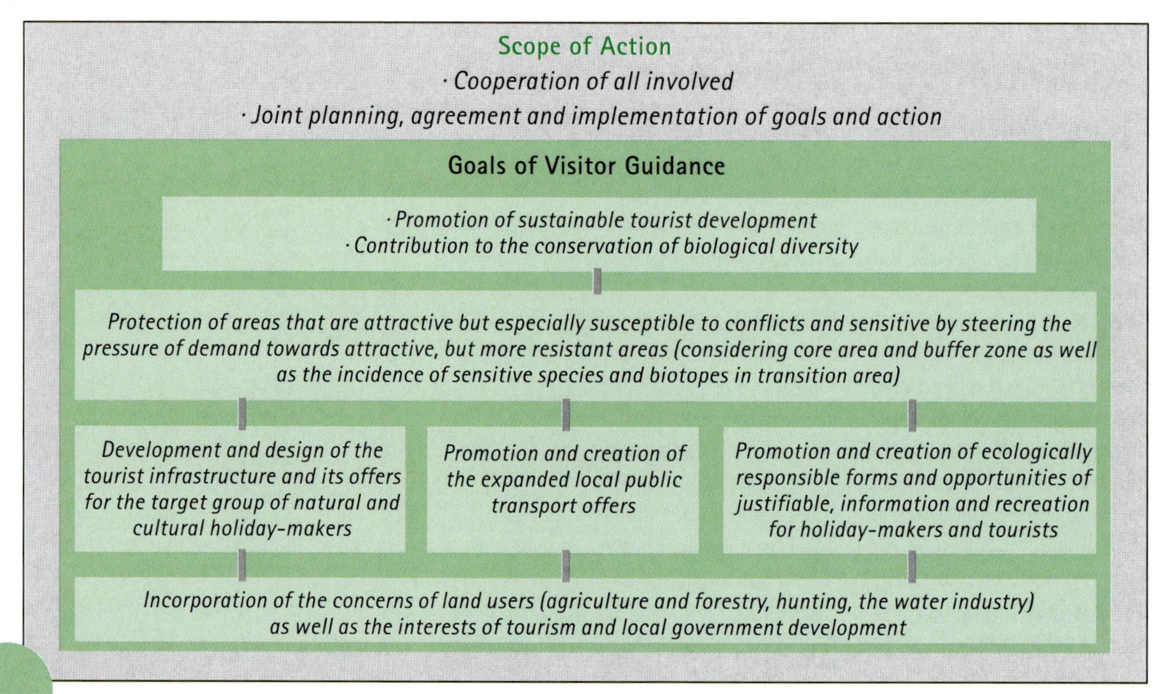

Fig.: Goals of visitor guidance (Revised from KLEINE-HERZBRUCH, N. 2000)

The findings of the undergraduate thesis also made clear that there was urgent need for action with regard to revising the tourist path network. The representatives of the region shared this view and the revision of the path network was agreed with them as the next part of the project.

Revising the Tourist Path Network – From Theory to Practice

A number of problems were identified when the starting situation was considered: there were conflicts in the field of nature conservation areas, with the concerns of forestry and hunting as well as in financing, e.g. path maintenance. Moreover, some visitors complained about the fact that there were several usages on the same path and about discrepancies between the signposts locally and the information on walkers' maps. In the next stage, a target for revising the network of tourist paths covering all interests was drawn up. Project-linked cooperation was agreed with the biggest landowner, *Thüringer Landesforstverwaltung* (Thuringian *Land* Forestry Administration).

This was followed by an identification of the specific problems in each municipality from the point of view of tourism, forestry and nature conservation.

In the main phase, "large scale" talks were held in the twelve towns and municipalities. Everyone who had any point of contact with the network of tourist paths took part in the discussion. In total, 80 consultations were held, involving the participation of over 70 representatives from various institutions. The BR administration took on a moderator's role for this part of the project and was responsible for implementing it.

In addition to the land in the Biosphere Reserve, this also dealt with areas in the surrounding Thuringian Forest Nature Park (a total of 36,680 ha). The BR administration used the "ArcView" geographical information system (GIS) in implementing this part of the project.

In August 2001, the revised network of paths was introduced and handed over to the public at a presentation ceremony (Hellmuth, E., Hörl, J. 2001). As a result, the network of tourist paths and, thus, the follow-on costs, were reduced and the tourist uses were simplified. For example, the total length of the ramblers' paths was shortened from 1,089 kilometres to 849 kilometres.

The paths largely cross state and local authority forests, with the planning taking account of route information systems and game introduction areas. The nature conservation areas, in particular the core zones, were calmed taking account of the incidence of species and biotopes susceptible to disturbance. Then, the municipalities evaluated the results of the revision. They estimated that the quality of the network of tourist paths will improve as a result of the restrictions and simplification of use.

The representatives from the region met in August 2002 and again in April 2003 to draw interim conclusions about the status of implementation. Implementation will probably take a long time.

Hiking tourists expect a clear network of paths. This was one of the reasons why the improvement of the network of paths in the Vessertal-Thuringian Forest BR was a prime objective of the "Visitor Guidance" project.

The following had been achieved by June 2003: the municipalities are updating the signposts. Some of the information boards have already been replaced. The *Thüringer Landesforst verwaltung* has included the revised network of tourist paths in its digital information system. Moreover, the forestry authorities have started to repair poor paths and to clear views. The BR administration has contacted publishers and authors of travel literature and has done the necessary groundwork. Various new editions of walkers' and ski route maps have already been published with the appropriate amendments. In June 2003 a local publisher in conjunction with the BR administration released the "Vessertal Biosphere Reserve" walkers' map for the first time.

The Basics of Nature Conservation – Questions and Answers

In her undergraduate thesis, Kleine-Herzbruch was able to make only an initial consideration of the aspects relating to nature conservation (Kleine-Herzbruch, N. 2000). Although the BR administration feels that this should have been dealt with in more detail, the revision of the network of tourist paths was started because the most urgent need for action was seen here. However, by the end of the year 2000 a contract for work and services to this effect was issued. Its subject was the development of geographically specified requirements for species and biotope conservation using visitor guidance in the Vessertal-Thuringian Forest BR (Opus 2002).

Starting from the research in the literature, lists of species and biotopes potentially susceptible to disturbance were drawn up. As well as their incidence in the area being studied, the Red List Status (Thuringian Red List) was another criterion used as a basis. Moreover, records were kept on which of these species are included in the Annexes of the EU Habitats Directive and in the leading and target species concept of the Vessertal-Thuringian Forest Biosphere Reserve (SCHLUMPRECHT, H. et al. 2002). In total, 55 animal and plant species and 20 biotope types potentially susceptible to disturbance have been identified.

The 55 animal and plant species susceptible to disturbance that have been identified include:

- Arnica *(Arnica montana)*,
- Broad-Leaved Marsh Orchid *(Dactylorhiza majalis)*,
- Elder-Flowered Orchid *(Dactylorhiza sambucina)*,
- Round-Leaved Sundew *(Drosera rotundifolia)*,
- Hare's Tail Cottongrass *(Eriophorum vaginatum)*,
- Siberian Iris *(Iris sibirica)*,
- Stag's-Horn Clubmoss *(Lycopodium clavatum)*,
- Early Purple Orchid *(Orchis mascula)*,
- Whinchat *(Saxicola rubetra)*,
- Tengmalm's Owl *(Aegolius funereus)*,
- Adder *(Vipera berus)*,
- Fire Salamander *(Salamandra salamandra)*,
- Brown Trout *(Salmo trutta fario)*.

The Fire Salamander (Salamandra salamandra) is one of the animal species most susceptible to disturbance in the Vessertal-Thuringian Forest BR.

However, for many species there is insufficient evidence of the actual impairments resulting from tourist activities. In most cases, other land uses, such as agriculture and forestry, cover up the damage from tourist activities, meaning that it is difficult to separate uses out in individual cases.

An evaluation with the maps showing the distribution of species and biotopes in the Vessertal-Thuringian Forest BR susceptible to disturbance drawn up with GIS showed a relatively even spread across the individual zones of the Biosphere Reserve. This means that aspects of visitor guidance must also be considered in the transition area.

As there are as yet no specific details on visitor behaviour from the Biosphere Reserve, the potential damage from tourist activities such as hiking, cycling, riding, skiing, nature photography, cross-country running, climbing, motocross, gathering mushrooms and berries as well as aviation sports were identified first of all. The possible damage is largely connected to people leaving the paths.

The results, in particular those of the research in the literature, prove that we do not yet have enough knowledge to answer important questions. For example, there are hardly any statements that can be generalised about the actual damage to species and biotopes resulting from tourist use. Neither is there a basis for the Vessertal-Thuringian Forest BR, in particular concerning further incidences of species potentially susceptible to disturbance, on the transferability of results from other regions and on visitor behaviour.

The Outlook – New Parts of the Project

In April 2003, the status of the implementation of the network of tourist paths, the results of the contract for work and services and its evaluation were presented to and discussed with the regional representatives.

The participants supported the proposal of the BR administration to set up a system of visitor monitoring in the Vessertal-Thuringian Forest BR to identify the actual damage to the species and biotopes potentially susceptible to disturbance.

There is further need for action resulting from the deterioration in local public transport offers for tourists and the car parking problems in winter. The municipalities have asked the BR administration for support in solving these problems.

Modern Participation Methods – The Key to Success

A major foundation for dealing with the project was the involvement as partners of the representatives from the region in the open dialogue. The intensive back-up and support for the project by the *Thüringer Landesforstverwaltung* also contributed to the success.

Various players from the region are involved in the "Visitor Guidance" project and regularly meet for consultations.

All of this was supported with intensive public relations work. Representatives of the press have repeatedly taken part in consultations and project discussions and regularly reported on them in the local press.
The presentation and discussion of the project at specialist conferences (Arnberger A. et al. 2002 and Moder F., Hellmuth, E. 2002) provided ideas for work and was registered positively in the region. Increasingly, the project description on the website www.biosphaerenreservat-vessertal.de is being used for information.

Thuringian Forest

The Right Framework Conditions – Positive Results

The people involved in the project constantly expressed their positive feelings about the course of the project and its results. Other factors alongside the transparent involvement procedure contributed to this assessment.
For various reasons there was a need for action with regard to the revision of the network of tourist paths and the partners in the region were therefore interested in solutions. On the whole, this part of the project was aimed at balancing interests, i.e. all partners acted with flexibility and a willingness to compromise. All involved underwent a learning process concerning the interests of their opposite partner.
Dealing with parts of projects also proved to be advantageous. This way of working was transparent because partial goals were formulated and partial results achieved. This became all the more important, the longer the project lasted.
It was also decisive that the BR administration had the requisite management capabilities to implement the project. The use of GIS in the BR administration meant that a modern system of data management was in use that made many evaluations much easier and made cartographic processing more time and cost effective.
An external expert was consulted for a few specific questions. This was where a need for further scientific studies became apparent.
At the same time, experience clearly showed that it was right to set the project in motion although there were many deficits, e.g. concerning the nature conservation foundations. Although these deficits still apply, the conclusion can be drawn that all those involved from the region have taken an important step towards sustainable tourism development. At this point they should be thanked for their dedicated work.

Literature

Arnberger, A. et al. (2002): Monitoring and Management of Visitor Flows in Recreational and Protected Areas. Conference Proceedings. January 30 – February 02 2002. Institute for Landscape Architecture and Landscape Management. Bodenkultur Universität Wien, Austria.
Hellmuth, E., Hörl, J. (2001): Projekt Besucherlenkung im Biosphärenreservat Vessertal. Teilprojekt: Überarbeitung und Abstimmung des touristischen Wegenetzes. Ergebnisbericht zur Überarbeitung und Abstimmung des touristischen Wegenetzes im Bereich der Forstämter Oberhof, Schmiedefeld und Ilmenau (Städte und Gemeinden im Ilm-Kreis sowie Suhl-Goldlauter). Verwaltung Biosphärenreservat Vessertal, Schmiedefeld a.R.: In: www.biosphaerenreservat-vessertal.de.
Kleine-Herzbruch, N. (2000): Ziele der Besucherlenkung im Biosphärenreservat Vessertal unter Berücksichtigung touristischer und naturschutzfachlicher Aspekte. Universität Gesamthochschule Kassel, Studienbereich Stadt- und Landschaftsarchitektur, Wintersemester 1999/2000. Unpublished thesis.
Moder, F., Hellmuth, E. (2002): Objectives and Basis of Management of Visitor Flows in the Biosphere Reserve Vessertal/Thuringia Germany. In: Monitoring and Management of Visitor Flows in Recreational and Protected Areas. Conference Proceedings. January 30 – February 02 2002. Institute for Landscape Architecture and Landscape Management. Bodenkultur Universität Wien, Austria, pp. 346-352.
Opus (Oekologische Studien, Umweltstudien und Service) (2002): Naturschutzfachliche Grundlagen der Besucherlenkung im Biosphärenreservat Vessertal (GIS-gestützte Analyse). Gutachten im Auftrag der Verwaltung Biosphärenreservat Vessertal, Schmiedefeld a.R. Unpublished.
Schlumprecht, H., Bock, K.-H., Erdtmann, J., Treß, J. & J. Wykowski (2002): Leit- und Zielartenkonzept für das Biosphärenreservat Vessertal. Gutachten Verwaltung Biosphärenreservat Vessertal, Schmiedefeld a.R. In: www.biosphaerenreservat-vessertal.de.

4.7 *Sustainable Agriculture on the Hallig Islands*

Schleswig-Holstein Wadden Sea Biosphere Reserve

Kirsten Boley-Fleet

Introduction

Off the North Sea coast of the *Land* Schleswig-Holstein in the middle of the Schleswig-Holstein Wadden Sea Biosphere Reserve and National Park lie the North Friesian Hallig islands. Their names are Gröde, Habel, Hamburger Hallig, Hooge, Langeness, Norderoog, Nordstrandischmoor, Oland, Süderoog and Südfall. Hallig islands are areas of salt marsh that are flooded sporadically during exceptionally high tides. They are an important component of the Wadden Sea ecosystem and, also, act as breakwaters to protect the coast from the storm tides of the North Sea.

The islands provide habitat for salt marsh plants and animals, e.g. Common Sea-Lavender (*Limonium vulgare*) and Sea Purslane (*Halimione portulacoides*) as well as unique weevils (e.g. *Apion limonii*) and spiders (e.g. *Erigone longipalpis*). They also offer breeding sites for many waders and waterfowl, e.g. Common Redshank (*Tringa totanus*), Oystercatcher (*Haematopus ostralegus*), Arctic Terns (*Sterna paradisaea*) and roosting and feeding sites for internationally important numbers of migrating birds, e.g. Brent Geese (*Branta bernicla*) and Knot (*Calidris canutus*).

Development of Agriculture on the Wadden Sea

For centuries people have lived on and farmed the so-called floating dreams (Stiftung Nordfriesische Halligen 2000). As protection against high tides the houses are built on man-made hills known as "*Warften*". Today just fewer than 300 people still live there. The people are firmly rooted in their homeland and have adapted to the specific challenges of life on the Hallig islands. Agriculture, coastal protection and tourism form the basis of their economy.

Until the mid-20th century agriculture provided the livelihood of the inhabitants; however, its importance has fallen steadily over the last few decades. The fact that the Hallig islands are still farmed at all, in spite of numerous adversities, is due to various funding schemes.

For example, in 1974 the Hallig islands were linked to the Mountain and Hill Farmers' Programme financed by EU funds and designed for disadvantaged areas throughout Europe (Directive 75/268/EEC 1975). The name of the programme is somewhat misleading, however. Outside mountainous and hilly regions farmers in disadvantaged areas with difficult climatic conditions and limited land use also received funding. Because of the specific problems associated with farming on the Hallig islands, the Schleswig-Holstein Ministry of Agriculture launched the Hallig Island Programme in 1987 (MELF 1986).

To this day the aim of the programme is to preserve the Hallig islands in their original, semi-natural character with their important ecological functions in the Wadden Sea and, at the same time, to secure them as a living and working environment for the indigenous population.

Since 1988 the European Community, now the European Union, has helped to finance this funding within the framework of co-financing. In 1992 the Hallig Island Programme was updated. More funding was made available in 1998 with the Programme for Contractual Nature Conservation in Schleswig-Holstein. The Hallig Island Programme is currently being revised again within the context of a plan by the *Land* Schleswig-Holstein to develop rural areas under EU legislation (Council Regulation (EC) No. 1257/1999 of 17 May 1999 (3)) and has been submitted to the European Union for notification.

The small but effective Hallig Island Programme has never been the subject of budgetary cuts in Schleswig-Holstein because there was political and social consensus that the Hallig islands with their traditional way of life and farming should be preserved.

The Hallig Island Programme

There are around 50 farms on the Hallig islands. They farm approximately 1,750 hectares of salt marshes. They are all forage-growing farms that have their own dairy cattle or that take in cattle from the mainland over the summer months.

Every spring and autumn the Hallig salt marshes provide valuable feeding areas for Brent Geese (*Branta bernicla*). However, these geese are a cause of concern to the farmers when they fly in to the Wadden Sea in the spring to feed on the Hallig salt marshes, so competing with domestic stock.

Who wouldn't want to spend the summer here? Cows on Nordstrandischmoor.

Approximately half of the biogeographical population of the Dark-bellied Brent Geese (*Branta bernicla bernicla*) utilise the Schleswig-Holstein Wadden Sea as a traditional stopover site on route between the Arctic and Europe. The *Land* Schleswig-Holstein has a special international obligation to protect this species. For the individual farmer, the grazing of Hallig salt marshes by the Brent Geese (*Branta bernicla*) is associated with considerable economic losses.
To preserve the ecological character of the Hallig islands and, at the same time, preserve the employment of the rural population, agriculture on the Hallig islands is subsidised. Contracts are made with farmers to recompense for nature conservation work, management restrictions and damage caused by Brent Geese (*Branta bernicla*).
Contracts can be made with farmers on the Hallig islands who keep cattle, sheep or horses there. The contracts run for a period of five years.

A Success Story with Good Prospects for the Future

The Hallig Island Programme is an exemplary success story. It was designed and implemented, and is constantly developed in cooperation with the local population. It combines ecological and economic interests. Even if there is occasional criticism – triggered by large numbers of geese or bad weather – the Hallig Island Programme is a good example of the successful cooperation of agriculture, nature conservation and local administration.
Among other things, this is indicated by the facts that:
- there is still farming on the Hallig islands;
- the total number of Brent Geese (*Branta bernicla*) recorded on the Hallig islands is stable;
- the inhabitants of the Hallig islands are actively marketing the success of the nature conservation activities (for example, the Hallig island communities of Langeness and Hooge organise "Brent Geese days" with a diverse range of events including guided tours and observation of Brent Geese) and
- every year there are Hallig islands assessment meetings where the Hallig island farmers and representatives from the competent authorities and stakeholders evaluate the effects of the programme on the Hallig salt marshes and discuss its amendment.

The Hallig Island Programme is part of the *Land* Schleswig-Holstein's plan to develop rural areas and will thus play an important role in the region in the future. Even now, an unforeseen positive development has come about: inspired by various information events, including those organised by the Biosphere Reserve administration, the inhabitants of the five big Hallig islands, which are not yet part of the Biosphere Reserve, have expressed a wish to join the existing Schleswig-Holstein Wadden Sea Biosphere Reserve as a transition area. They are not only convinced that their traditional sustainable farming fits into the UNESCO Man and the Biosphere Programme (MAB), but also want to preserve their living and working environments, secure their livelihoods and open up perspectives for their children.
UNESCO recognition is not a guarantee for generous funding programmes; however, it can turn out to be an advantage in the national, European or worldwide competition for grants. There is a already a little "reward": the Uthlande region, to which the Hallig islands belong, has successfully taken part in the nationwide *Regionen Aktiv* competition organised by the Federal Ministry for Consumer Protection, Food and Agriculture (BMVEL). The projects for regional marketing, promoting and improving sustainable tourism on the Hallig islands and environmental education concepts designed by the inhabitants of the region can now be realised. This will greatly advance development on the Hallig islands. Present and planned projects on the Hallig islands demonstrate the innovative nature of the agricultural communities living there. They are well on their way to a sustainable future.

Literature

SCHWABE, M. (2000): Das "Halligprogramm" des Landes Schleswig-Holstein.
PROKOSCH, P. (1989): Ringelgänse wieder am Pranger. In: Wattenmeer International, June 1989.
MINISTER FÜR ERNÄHRUNG, LANDWIRTSCHAFT UND FORSTEN DES LANDES SCHLESWIG-HOLSTEIN (1986): Halligprogramm.
BANCK, C. (2000): Manuskript über ein geplantes Buch über die Halligen. Unpublished.
MINISTER FÜR NATUR, UMWELT UND LANDESENTWICKLUNG (1992): Richtlinie für die Gewährung eines erweiterten Pflegeentgeltes sowie einer Prämie für natürlich belassene Salzwiesen in Anlehnung an das Halligprogramm. Bekanntmachung des MNUL vom 10.03.1992, XI 530/3217.6600 (Amtsblatt Schleswig-Holstein, S. 213), berichtigt am 27.04.1992, XI 530/5327.6600 (Amtsblatt Schleswig-Holstein, p. 310).
PLANUNGSBÜRO PRO REGIONE (2001): Jahresbericht 2001 zur Untersuchung der Salzwiesen-Brachen.
LANDESAMT FÜR DEN NATIONALPARK SCHLESWIG-HOLSTEINISCHES WATTENMEER (1998): Das Halligprogramm. In: Nationalpark Nachrichten 5/98.
LANDESAMT FÜR DEN NATIONALPARK SCHLESWIG-HOLSTEINISCHES WATTENMEER (1998): Fraßschäden durch Enten und Gänse. In: Nationalpark Nachrichten 3-4/98.
STIFTUNG NORDFRIESISCHE HALLIGEN (2000): Schwimmende Träume.
RICHTLINIE 75/268/EWG vom 28.04.1975 des Rates über die Landwirtschaft in Berggebieten und in bestimmten benachteiligten Gebieten. Abl. L 128 vom 10.05.1975.

4.8 Environmental Education: A Component of Sustainable Development

Upper Lausitz Heath and Pond Landscape Biosphere Reserve

Peter Heyne

Environmental education should meet the following requirements under the Criteria for Designation and Evaluation of UNESCO Biosphere Reserves in Germany:

- improvement of environmental know-how, and development of well-founded environmental knowledge,
- providing an opportunity for direct encounters with the naturally and anthropogenically shaped environment, and for recognising and evaluating factors that influence this environment;
- providing an opportunity for study of, and reflection about: the current environmental situation and its history, and the relationships between human beings, their social institutions and their naturally and anthropogenically shaped environment;
- development and teaching of alternatives to behaviour and attitudes that have been recognised as environmentally harmful.

  (German MAB National Commitee 1996; National Criteria cf Annex, cf p. 164)

The Aims of Environmental Education

Environmental education is aimed at understanding the relationship between humans and the environment, character formation, values and behaviour. It requires an intensive dialogue with smaller groups and thus reaches a smaller number of people in comparison to public relations (PASCHKOWSKI, A. 1996). Successful environmental education requires educational concepts tailored to the individual target groups and a continuous way of working.

Environmental education should lead to identification with the (home) region, portray nature and mature cultural landscapes as something beautiful and aesthetic, worthy of respect and protection; it should address all of the senses and demonstrate approaches and examples of how to treat nature with care.

Environmental education must realise a holistic approach and teach social skills as well as experience with and of nature. Environmental education work should always be related to the current problems in the area, but should not ignore global environmental problems.

Environmental education should teach how sustainability can be achieved in the area. Other subjects should be the particular features of the landscape, e.g. for the Upper Lausitz Heath and Pond Landscape Biosphere Reserve, ponds, wetlands and arid regions in close proximity to each other, dunes and post-mining landscapes. Furthermore, environmental education should deal with cultural and social characteristics of the region: in the Upper Lausitz Heath and Pond Landscape BR, the special issues include: the history of settlement, the life of the Sorbs – a small national minority in the Lausitz – or rural unemployment, which also greatly affects the region. The result of environmental education should be responsible action towards one's fellow people and the environment.

Environmental education in the Upper Lausitz Heath and Pond Landscape BR concentrates on the "local players", i.e. the people who live and work here, to take responsibility for the goals of the Biosphere Reserve in all areas of life and to implement appropriate measures. Environmental education should integrate the cultural community of the village, come to terms with it and enrich it.

Environmental education must aim to reach all age groups. The work with children should lay foundations and help them to develop their values. Adult education should arouse curiosity, convey pleasure, encourage understanding of traditions, deepen or revive knowledge, identify problems and help to find a solution. Finding allies for the education of children and young people and finding common projects is a major goal of educational work for adults (BIOSPHÄRENRESERVATSVERWALTUNG 2000).

To achieve these goals, we offer many different events and campaigns in our Biosphere Reserve that use specific environmental education methods and are constantly further developed.

Offers in the Upper Lausitz Heath and Pond Landscape BR

Environmental education in the Upper Lausitz Heath and Pond Landscape BR is primarily aimed at children and young people, but also at adults.

With an offer that is as specific as possible, we want to reach very different target groups with very different expectations. We can organise an extensive offer with two permanent employees and seasonal freelance staff.

Our range for children and young people includes nature experience walks, programmes in the classroom, project days, competition or holiday camps tailored to subjects and ages.

Examples of holiday camps:
- "Faces of the Landscape – Artistic, Natural History Camp"
- "Stork Journey Holiday Camp – A Bicycle Tour in Search of the Stork Through the Biosphere Reserve"
- "Hunters, Gatherers, Cultivators – On the Tracks of an Old Cultural Landscape"
- "Otter Camp – In the Tracks of the Lausitz Aquatic Mammal"

It is also especially important for us to establish and regularly care for spare time groups. In these time groups the children can take part in fun projects, e.g. looking at bats and potato farming and, in the process, making an important contribution to protecting nature.

Project day: Children are the most important target group of environmental education

To take the issues of environmental education into the families we organise small family parties together with the spare time groups. Here, the children that we look after in the spare time groups may present short plays to their parents or, in the "autumnal potato festival", offer them produce that they have grown themselves.

There is a puppet show for environmental education in larger groups. On the basis of a familiar, updated German fairy story it deals with the subject of "carp production", which is very important in the region.

For adults, we have to take account of a greater dispersal of the target group. Lectures, seminars, excursions, colloquia and various competitions are offered in order to reach interest groups that are as diverse as possible.

Every year in our Biosphere Reserve we organise approximately 260 excursions, 30 project days, 30 nature experience walks, five holiday camps, 80 spare time group meetings, four family parties, six seminars and two colloquia.

The successful work in environmental education has been very important in the acceptance of the Biosphere Reserve among the local population. In a survey in 2002, 76 per cent of those questioned believed that the Biosphere Reserve was a "sensible facility", only three people rejected it (Brassel, V. 2002).

Do You Like Bats?

One example of particularly successful spare time group work was the project "Do You Like Bats?". Within a previous project week at a middle school, 10-year-old pupils were set the task of researching the distribution of bats in their community. Before small groups of children with questionnaires interviewed the village residents, they were prepared in working groups. They composed short talks on the lives of the animal group and presented the results to each other. Games lightened up this theoretical part. Various questioning situations were acted out: first of all a sociable resident followed by a more unfriendly compatriot. Strengthened by this training, equipped with questionnaires and detailed maps of the locality, the children set out on their way. On the fourth day the results were displayed on large boards and presented at school.

The main aspect of village mapping was the personal contact between the children and the village residents. After all, it takes some courage to ring at every door and ask about bats. On the other hand, the adults were mainly open to the serious questions from the children.

This project week led to a spare time group with an interested group of pupils, who then concerned themselves with practical measures for bat protection for a whole year. The pupils mapped tree hollows, attached nesting aids to houses and built a pond to improve the food on offer. The bat boards still adorn some facades, and the attention that many inhabitants pay to the "pixies of the night" in this village has certainly changed for good.

Literature

Biosphärenreservatsverwaltung (Ed.) (2000): Biosphärenreservatsplan Teil 2. Rahmenkonzept für Schutz, Pflege und Entwicklung, Mücka.

Brassel, V. (2002): Analyse der Wirksamkeit der Öffentlichkeit im Biosphärenreservat "Oberlausitzer Heide- und Teichlandschaft". Diplomarbeit. Albert-Ludwigs-Universität, Freiburg.

Dietz, M., Meißner, T. (2000): Kinder der Dörfer. Umweltbildung im Biosphärenreservat Obere Lausitz Heide und Teichlandschaft, Bautzen.

German MAB National Commitee (Ed.) (1996): Criteria for Designation and Evaluation of UNESCO Biosphere Reserves in Germany, Bonn.

Paschkowski, A. (1996): Rahmenkonzept Umweltbildung in Großschutzgebieten. WWF-Naturschutzstelle Ost, Potsdam.

4.9 Health and the Biosphere Reserve

Berchtesgaden Biosphere Reserve

Werner d'Oleire-Oltmanns and Ulrich Brendel

Health as a Guiding Principle for Sustainable Development

The Berchtesgaden UNESCO Biosphere Reserve differs from the other biosphere reserves in Germany mainly in terms of its topographical conditions. This is especially clear in a well-defined north-south gradient of over 2,000 metres between Bad Reichenhall at 470 metres above sea level and the summit of the Watzmann at 2,713 metres.

Typical landscape in the Berchtesgaden Biosphere Reserve

This is the reason for a diverse landscape with varied habitats. Topographical conditions and versatile habitats have also greatly influenced the intensity of usage and led to a small farm landscape structure that is extremely rich in species.

As well as intact nature and centuries of sustainable use, health is to shape the future guiding principle in the regional tourism sector. To make optimal use of the nature potential in the region, the healthy mountain climate must be brought well to the fore when marketing the region. Against the background of an increasing number of people suffering from allergies, a new direction appears to make sense.

Favoured by the Climate

The special climatic conditions in the Berchtesgaden BR are ideally suited to a facility of this kind. It lies in the transitional area between an oceanic and a continental climate and also has a typical mountain climate due to the height difference of 2,000 metres.

Various parameters such as the height, exposure and slope gradient are responsible for the great vertical, horizontal and seasonal variability of the Biosphere Reserve. Furthermore, the practically linear characteristic of the annual mean temperature and precipitation ensure a low-allergen climate in the recognised pure air area.

All of these regional advantages have been recognised by the *Deutsche Gesellschaft für Klimatherapie e. V.* (German Society for Climate Therapy) in Berchtesgaden. This body should be consulted when creating a scientifically and medically reliable evaluation of accommodation with regard to their suitability (certification) for people with sensitive reactions.

Geographic Information Systems and the Biosphere

Within the context of the project "Berchtesgaden Ecosystem Research" of the UNESCO Man and the Biosphere Programme (MAB), habitat models were developed to protect the Golden Eagle (*Aquila chrysaetos*) in the Alps (Brendel, U. et al. 2000). The same method, which uses Geographic Information Systems (GIS), is used for house dust mites (*Dermatophagoides pteronyssinus and Dermatophagoides farinae*) or mould fungi (e.g. *Aspergillus* spp., *Mucor spp. or Cladosporium* spp.). The phenology of pollen flight can also be realistically presented with the help of GIS. The data collected can be extrapolated to larger areas and examined there with appropriate indicators.

Such an extrapolation of methods precisely matches the concept of the UNESCO biosphere reserves: research, develop, try out, apply locally and publicize the results and experience in the region and beyond it to make them usable.

The first stage of implementation is to draw up a three-stage certification system to classify the allergen contamination for businesses (hotels, restaurants, etc.) as well as places in the open air within the Biosphere Reserve and to document them in a map.

Mould funghi

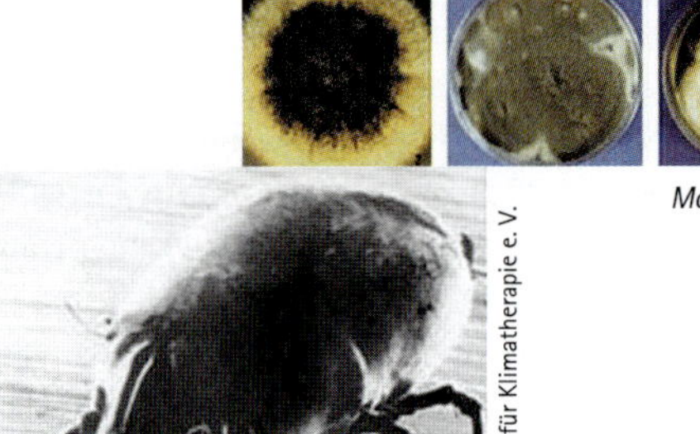

House dust mite

In a second stage, the models are extended to allergens such as pollen and air pollutants (ozone, nitrogen oxide, sulphur dioxide, fine dusts) and, again, represented in map form with statements on the geographic distribution of pollution.

Holiday without Allergies

On the basis of the results, an allergy information system for patients is to be set up as a further focus. It can be used to recruit tourism offers for a low-allergen or allergen-free holiday.

The allergy information system is to be linked to the www.info-bgl.de information platform that is currently being developed within the framework of an EU project. All of the means that a user needs to plan his or her holiday (searching for restaurants, bus and train timetable information, etc.) are available for this.

A combined information system of this kind would be another unique selling point for the Berchtesgaden BR. It would thus not only make it stand out from all of the German biosphere reserves, but also from all other health regions in the Alps.

Literature

Brendel, U., Eberhardt, R., Wiesmann-Eberhardt, K. & W. d'Oleire-Oltmanns (2000): Der Leitfaden zum Schutz des Steinadlers Aquila chrysaetos (L.) in den Alpen. Nationalparkverwaltung Berchtesgaden, Forschungsbericht Nr. 45.

4.10 *Natural Dynamics at the Heart of Europe*

Lower Saxon Wadden Sea Biosphere Reserve

Irmgard Remmers

One of the Last Natural Landscapes

Natural, unused or little used ecosystems are rare in Central Europe. In its large core area and buffer zone the Lower Saxon Wadden Sea Biosphere Reserve protects just such an ecosystem: the Wadden Sea, one of the last natural landscapes whose key characteristic is its high natural dynamism.

The example of the Lower Saxon Wadden Sea shows that a natural landscape still has its place in densely populated Central Europe and what form possibilities for the coexistence of people and nature can take.

The Sea Floor Meets the Horizon

The Lower Saxon Wadden Sea Biosphere Reserve covers around 240,000 hectares between the Ems and Elbe estuaries. It comprises practically all of the Lower Saxon Wadden Sea with its extensive mudflat and water areas, the foreshore in front of the dykes and the East Friesian Islands with their dune and salt meadow habitats.

The core area and buffer zone lie almost completely in the approximately 280,000 hectare national park of the same name. The changing tides and the resulting great natural dynamism are the elements that shape the landscape in the Biosphere Reserve: twice a day around 220,000 hectares of mudflats fall dry and then are flooded again.

Pure dynamism:
mudflat and tidal inlet system in the core area of the Biosphere Reserve

The high conservation value of this landscape is not only demonstrated in its designation as a UNESCO biosphere reserve and a national park, the Wadden Sea is also a conservation area according to the EC Birds directive and registered as an area according to the EC Flora-Fauna-Habitats-Directive. An application for recognition as a World Natural Heritage Site is also ongoing.

Humans and Natural Dynamism

The coastal area has been a place of human settlement and farming since time immemorial. Over the centuries the great dynamism and the dominant natural forces with storm tides, coastal erosion and island movements have forced use adapted to the natural conditions. This changed with the technical possibilities of the twentieth century.

It is the job of the administration of the conservation area to guide and weigh up the various interests to enable coexistence between the people who live, work and seek recreation in the coastal area and the nature conservation objective of "preserving the great dynamism that is characteristic of the Wadden Sea".

Holiday in a Biosphere Reserve – Naturally!

Every year over two million holiday-makers visit the southern North Sea holiday region with its seven islands and the many coastal resorts. Whereas the accommodation lies outside the Biosphere Reserve, the majority of tourist activities are concentrated on the beaches located in the transition area of the Biosphere Reserve. But the holiday-makers' activities, such as walking, cycling and riding, also take place on the paths in the buffer zone and core area.

The spatial distinction between the strictly protected core area, the equally heavily protected buffer zone and the transition area that is available for bathing and spa treatments without restrictions, means that sensitive coexistence between holiday-makers and nature is possible. A concept for paths ensures that it is possible to experience the many facets of nature and landscape even in the core area without making any disturbance into the area.

Jürn Bunje, 2001; "Lower Saxon Wadden Sea" BR

Viewing platform in the core area at the eastern end of Langeoog

As part of the public relation and educational work, visitors are informed about information material and, in the 14 National Park houses and centres, about the habitat and its special conservation value. In the area itself a system of protected area wardens looks after the visitors.

Jürn Bunje, 2003; Lower Saxon Wadden Sea BR

Guarantee for recreation: holiday in a biosphere reserve. Bathing beach in the transition area

Semi-Natural Development of Salt Marshes

Non-extension of lease agreements on land belonging to the regional government, the voluntary surrender of use and land sales mean that 66 per cent of the salt marshes in the core area and buffer zone of the Biosphere Reserve are not used. They are left for semi-natural development. This nature conservation objective was achieved in harmony with the concerns of coastal protection and taking account of the economic impact on farmers. Extensification adapted to the concerns of nature conservation and specific species conservation has been achieved for 23 per cent of the salt marshes.

Jürn Bunje, 2000; Lower Saxon Wadden Sea BR

Habitat between the land and the sea; Natural salt marshes on the eastern coast of the Spiekeroog island in the core area of the Biosphere Reserve

As Much Coastal Protection as Necessary – as Much Nature Conservation as Possible

This fundamental maxim applies to all coastal protection measures in the Biosphere Reserve to harmonise the justified interests of coastal protection and nature conservation.

Numerous agreements to take greater account of nature conservation concerns in all coastal protection measures have been made by means of *Land* government requirements, foreshore management plans and detailed consultations on site.

Fishing – Traditional Use of the Sea with a Future

The shrimp fishing (fishing for prawns, *Crangon crangon*) done in the Biosphere Reserve with small cutters seems to be irrevocably linked with the traditional image of the Wadden Sea. Even today, this traditional form of use has its place in the Biosphere Reserve. To reduce the impact on the habitat, the administration of the conservation area supports measures to improve the catching and sorting techniques.
In mussel fishing (fishing for *Mytilus edule*) the use or non-use of existing mussel banks is being jointly regulated by the fishing industry and the management so that this form of fishing can continue in the Biosphere Reserve in the long run.

What would a holiday on the North Sea be without a cutter harbour and shrimp sandwiches? Traditional usages still have their place in the Biosphere Reserve.

The Wadden Sea as Part of the Biosphere as a Whole

Natural habitats in densely populated Europe can only ever be islands in the human economic area. This is especially the case for such open system as the marine conservation areas. Many of the external effects, such as pollutant inputs from rivers, the atmosphere and the high sea cannot be directly influenced by the administration of the conservation area. Nevertheless, it is possible to conserve a natural dynamism within the borders of a large conservation area.

This is where the Biosphere Reserve in conjunction with the National Park makes an important contribution because plans and measures that could have negative consequences for the conservation area should not be considered, or only in exceptional circumstances, in the actual area.

Untamed Power of Nature at the Heart of Europe

In spite of the usages in and on the margins of the Biosphere Reserve, humans have not succeeded in taming the natural forces of the sea and the tides. Enormous mudflat and water areas, dunes and salt marsh landscapes are still completely left to themselves and natural dynamism.
The example of the Lower Saxon Wadden Sea shows that it is possible to conserve a habitat of this kind and, at the same time, all usages that do not damage the overall ecosystem. On the contrary, the Biosphere Reserve is an invitation to an encounter with nature and – where this is wanted – to manage it in an environmentally sound way.
So there is still such a thing as natural, dynamic habitats at the heart of Europe. We are obliged to conserve them. Biosphere reserves, together with the other protection categories, make an important contribution to this.

*Where the wind defines the shape:
white dunes in the core area of the Biosphere Reserve*

4.11 Management of Migrant Birds

Elbe River Landscape Biosphere Reserve

Brigitte Königstedt

Elbe Valley Meadows – a Resting Centre

With a maximum of 45,000 White-Fronted Geese (*Anser albifrons*) and 25,000 Bean Geese (*Anser fabalis*) and 1,500 Mute Swans (*Cygnus olor*), 2,500 Whooper Swans (*Cygnus cygnus*) and 3,000 Bewick's Swans (*Cygnus columbianus*) the Lower Saxon section of the Elbe River Landscape Biosphere Reserve, the Lower Saxon Elbe Valley Meadows, is the most significant resting and wintering area for the above-mentioned species (SÜDBECK, P., KÖNIGSTEDT, B. 1999). Consequently, an area of 325 square kilometres has been identified as the Lower Saxon Middle Elbe EU Bird Conservation Area.

A wintering Whooper Swan family (Cygnus cygnus) in the Elbe valley meadows

In line with the trend across Europe, it can be assumed that the populations of resting Nordic swans and geese – Whooper Swans (*Cygnus cygnus*), Bewick's Swans (*Cygnus columbianus*), White-Fronted Geese (*Anser albifrons*) and Bean Geese (*Anser fabalis*) – have increased along the Elbe in the past decades. Censuses taken since 1995 indicate that the populations now remain stable at a high level. Despite generally good resting conditions, the birds are also exposed to dangers and disturbances here. In addition to the 'wiring up' of the landscape with overhead electric cables, endangering factors mainly include deliberate disturbances caused by people, such as the active chase or hunting. This occurs particularly within agriculture, where competition between the users of the land and the birds is a cause for concern.

Damage to Crops – Conflicts with Agriculture

The resting conditions for Nordic geese and swans have improved in the last four decades due to the development of the Elbe marshland as an extensive agricultural landscape and the vast cultivation of Winter Rape (*Brassica napus*) and grain, such as Winter Wheat (*Triticum aestivum*) and Barley (*Hordeum vulgaris*). Large groups of up to 6,000 migrant birds on farming land can lead to considerable drops in yields. The importance placed on the international responsibility for the protection of migrant birds can lead to a high potential for conflict at a regional level between agriculture and the government's conservation goals.

Migrant Bird Management through Contractual Nature Conservation

The Lower Saxon government's goal is to avoid compensation payments and use contractual nature conservation as a solution. The pilot project "Grazing Areas for Migrant Birds in the Elbe River Landscape" from 1994 to 1999 created fundamental specialist foundations. The following management measures were used:

- Provision of attractive food rich in energy (resting areas) and feeding areas free of disturbances (tolerance areas) through farming agreements.
- Reduction in hunting on project and sleeping areas through voluntary agreements with game tenants.

It has been proved that resting areas with ripe grain and maize result in geese gathering together locally for a short time. The tolerance areas for swans and geese are also used more intensively and, despite the high migrant bird numbers, the real drops in yields have been astonishingly low.

The "Grazing Areas for Migrant Birds in the Elbe River Landscape" project was implemented within the "Conservation of Biological Diversity – Nordic Migrant Birds" Cooperation Programme as part of the five-year "PROLAND LOWER SAXONY" promotion programme. Within this programme, contracts for 270 hectares of land were concluded in 2001. In addition, 500 hectares in 2002 and 960 hectares in 2003 were contractually designated as areas of tolerance for migrant birds on winter rape.

Additional measures include effective regulations for the hunting of game birds on sleeping areas and further agreements exist with game tenants. As part of routine actions taken to restructure the agricultural land of a community, it is envisaged that suitable resting areas will be provided for birds and that the number of overhead cables will be reduced.

In-depth management of migrant birds in the Lower Saxon Elbe Valley Meadows contributes to minimising conflicts between agriculture and conservation and to preserving the importance of the Elbe Valley Meadows as a resting area for birds.

Literature

SÜDBECK, P., KÖNIGSTEDT, B. (1999): Gänseschadensmanagement in Niedersachsen. In: NNA Ber. 12 (3).

4.12 Traditional Farms and the Spree Forest Landscape

Spree Forest Biosphere Reserve

Michael Petschick and Christiane Schulz

Introduction

The Spree Forest Biosphere Reserve lies in the eastern German *Land* of Brandenburg near Berlin. This biosphere reserve contains a model case of the "greening" of agriculture.

The picture of a "mosaic" landscape is still considered to be typical of the entire Spree Forest. But in fact there are only a few farms left that still apply traditional land use systems on a large scale. They can be seen most in the villages of Lehde and Leipe in the inner Upper Spree Forest. Since 1992 the administration of the Spree Forest Biosphere Reserve has been intensively involved with various partners to conserve these typical usage structures.

Island of fields in Leipe

Spree Forest BR Archive

Ways Towards Sustainable Development

When the two German states were unified in 1990 the instruments of European agricultural policy also started to have a direct impact in the Spree Forest. Under the changed agricultural policy conditions, the islands surrounded by a labyrinth of "Fliesse" (the canals typical of the Spree Forest) with grassland use or man-made ridge beds (pocket fields) are completely unprofitable. Alone the costs that have to be borne by a farm that can only be accessed with a Spree Forest boat on narrow canals are well above the revenue from growing vegetables or from selling meat from the cattle fattened here. Left to their own devices to a certain extent, many farmers wanted to give up in the early 1990s.

Here is an example of a typical farm:

Lehde

The farm has been family-owned for around 200 years and today it is farmed all year round by a person in the 41-50 age category. Until 1929 the farm could be accessed only by boat and even today water borders three quarters of the land, which can be accessed by land only via a narrow drive.

Arable land (0.5 hectares) and pastureland (8.0 hectares) have to be farmed with the Spree Forest boat as a means of transport. Winter fodder is stored in hay stacks in the fields. Due to a lack of space on the island there is no barn. Ten cattle and two pigs are the livestock on the farm.

The farmer's technology is largely old-fashioned because most of the work has to be done manually. On the farm island itself, attention is immediately drawn to a lovingly tended cottage garden, farmyard trees and greenery around the house.

The decree designating the Spree Forest Biosphere Reserve forms the basis of a joint concept to conserve these farms. The Law makes species diversity and the landscape that attracts four million tourist every year (surveys by *Tourismusverband Spreewald e. V.* [Spree Forest Tourist Association] 2002) into commodities of the UNESCO Biosphere Reserve that are especially worthy of protection.

Traditional Spree Forest farm in Lehde

Spree Forest BR Archive

Consequences of Dying Farms

The departure of agriculture from the land has serious consequences because it goes hand in hand with the loss of the unique cultural landscape. The dependency between the landscape and agriculture very soon became apparent to locals, farmers, the local authorities, the tourist industry and the administration of the Biosphere Reserve. The departing farmer leaves the land at the mercy of natural succession. This results in scrubbing over and forest formation. The original diversity of habitats for animals and plants and, ultimately, the entire landscape will be lost.

The farm is threatened with economic, the farmer with social decline. The traditional knowledge and the wealth of experience of the Spree Forest farmers in handling the landscape and the associated ways of farming it are at risk of being lost for ever.

Succession on cropland: the forest is returning

Strategies for Sustainable Development

For as long as the former German Democratic Republic existed (until 1989/90) the farmers in the Spree Forest primarily produced food. The cultural landscape resulted from this and was, to a certain extent, a – not highly respected – "by-product". Today, in 2003, his main task is to shape the cultural landscape. To conserve farms in this situation, farmers and other land users, such as tourists, first of all have to develop a shared awareness of this new role.

By founding local associations the farmers affected have themselves taken the first step in this direction and sought acceptance. This initiative has been supplemented by active partnerships among all groups interested in the conservation and further development of the Spree Forest farms. Since 1992 the Lehde/Leipe Working Group has constantly raised project funding of approximately € 300,000 per year and has used this money directly on the farms as reimbursement for its services for measures to conserve the landscape by means of traditional farming methods. Since the year 2000 an EU funding directive (EC No. 1257/99) has also been used to support small-scale arable farming.

In the long run, however, the farms should become self sufficient. For example, activities in direct marketing, the cultivation of old crop varieties or using the farms for tourist offers can contribute to achieving this goal. For the future it will be important for farmers to use the entire farm, their products and their traditional knowledge.

Goals for the Future

The farmers' partners all agree that the Spree Forest farms shape the identity of their home region and must be conserved under all circumstances. Together, all affected (the farmers, local authorities, tourism association, Spree Forest association and Biosphere Reserve administration) want to show the way for the next generation.

For this, the following subject areas must be dealt with:

- multifunctional agriculture – what values does society recognise;
- the landscape as a product and infrastructure;
- the Spree Forest farm as a venue for environmental education;
- new technologies and old crop varieties;
- marketing strategies with the pan-European protected geographic label "Spree Forest";
- scheduled generational change and future for young farmers;
- EU subsidies, agricultural environment programme, contractual nature conservation, regional activities;
- financial securing of projects by a Spree Forest Foundation that is yet to be founded.

Future for the Spree Forest farmers: the traditional way of farming contributes to the conservation of the cultural landscape.

These approaches can be implemented only if the regional players continue to cooperate effectively. This cooperation is not a one-off gift; it must always be made tangible in recurrent detailed work with all involved. The Biosphere Reserve administration takes on a coordinating function here, becomes involved as an originator of ideas and information platform for regional and supra-regional contacts and even secures some financial grants. However, the farmers themselves bear the responsibility for this process.

4.13 Cooperation between the German and Chinese MAB National Committees

Jürgen Nauber and HAN Nianyong

Within the framework of the UNESCO Man and the Biosphere Programme (MAB), cooperation between Germany and China began in 1987. At that time, the German Federal Ministry for Research and Technology (BMFT) financed the Cooperative Ecological Research Project (CERP) to the tune of US$ 4.8 million, made available to UNESCO as trust funds.

These funds financed eight interdisciplinary research sub-projects that were carried out in China from 1987 to 1995. The ecological problems China was experiencing due to its high population growth and rapid economic development determined the subjects for investigation. The main problems include water pollution, soil erosion, deforestation and degradation of ecosystems. UNESCO published the results of the projects in 1996 (UNESCO 1996).

Unlike in Germany, where the responsibility for the MAB Programme was transferred from the Federal Foreign Office to the Federal Ministry for the Environment, Nature Conservation and Nuclear Safety (BMU), in China the "Academy of the Sciences", the umbrella organisation of all the scientific institutions in the country, is the contact partner for UNESCO. The Chinese and German cooperation therefore also focused mainly on science at the beginning, which worked considerably to the advantage of the research carried out at this time. However, as China's scientific institutions had only minimal influence on the actual management of the land, the results were often not implemented or used. This explains why the Chinese "Academy of the Sciences" has attached increasing importance over time to the UNESCO biosphere reserves in the country for the implementation of research results. They are also using them increasingly as an instrument for land use planning for sustainable regional development. Accordingly, within the framework of the MAB Programme, Chinese and German cooperation is concentrating more and more on biosphere reserves. Mutual visits of the MAB National Committees have been arranged to understand the situation in each other's country and to begin a mutual process of exchange.

In 2001 a Chinese delegation, consisting of members of the Chinese National Committee and managers of biosphere reserves, visited Germany. In 2002 a German delegation paid a return visit to China. The Chinese MAB National Committee (www.china-mab.org) currently consists of 47 members from very different scientific disciplines and various positions of the government. It is extremely active in using the concept of biosphere reserves as a model solution for ecological problems and for sustainable regional development. In addition, the Chinese National Committee produces a monthly newsletter in English. It also publishes reports from the biosphere reserves and research results with impressive photographs in a Chinese magazine.

The biosphere reserves in China

The Chinese MAB Delegation in Germany

From 2 to 11 October 2001 a total of ten Chinese guests visited the Rhön and Schorfheide-Chorin Biosphere Reserves and the Federal capital city Berlin.

The Chinese MAB delegation in the Rhön Biosphere Reserve

The focus of the study trip was not the conservation aspect of biosphere reserves but the contribution that biosphere reserves can make to sustainable regional development.

With experts on site, the guests discussed issues relating to the construction of regional production cycles, tourism and environmental education. With local contributors, the administration and non-government organisations, they discussed the organisation required for the participation of the local population. Together with German experts, the participants discussed at the final workshop how biosphere reserves can contribute to the implementation of the

Convention on Biological Diversity (CBD). The biosphere reserves have strengths in the areas of conservation and sustainable use of biological diversity in particular. In addition, the visitors discussed with their German hosts cooperation within the framework of the "Clearing House Mechanism", the information platform for the application of the CBD, where the data and experiences available to the biosphere reserves can be made accessible worldwide.
The result of the visit was summarised by Mr CAO Guangzhao, Director of the Nanji Biosphere Reserve, as follows: "In Germany there is good cooperation between the administration and various conservation groups and environmental education works very well. The local economy can benefit from important incentives from a biosphere reserve with environmentally-friendly hotels and organic farming. The combination of conservation and organic farming contributes to the local sustainability."
Mr ZHAO Xiadong, Director of the Gaoligong Biosphere Reserve, acknowledged that although German experiences could not be directly applicable to China, he had gained important ideas for his work.

German MAB Delegation in China

Representatives of the German MAB National Committee and German biosphere reserves paid a return visit to their Chinese partners in September 2002. The German delegation visited the Huanglong, Juizhaigou and Wolong Biosphere Reserves in Sichuan Province and participated in a final workshop in the "Academy of the Sciences" in Beijing.

The German Delegation in the Juizhaigou Biosphere Reserve

China currently has 24 biosphere reserves, which since 1993 have been combined to form the Chinese Biosphere Reserve Network (CBRN). Together with the Democratic Republic of Korea, Japan, Mongolia, the Russian Federation and the People's Republic of Korea, they form the East Asian Biosphere Reserve Network (EABRN), which includes about 40 biosphere reserves.
China, too, is increasingly using the biosphere reserves to establish sustainable regional development and at the same time to preserve natural, precious features. At a meeting of the Chinese biosphere reserves in April 2003, the further development of Chinese biosphere reserves and current problems were addressed (Chinese MAB National Committee 2003). As is the case in many other countries, the dividing into zones of the Chinese UNESCO biosphere reserves must meet the requirements modified in the Seville Strategy: core areas that are too extensive sometimes can prevent economic activity. That makes it more difficult for the population to accept, which is contrary to the concept behind the MAB Programme. A balance must be negotiated once again between the interests of use and those of conservation.
Another issue much discussed in China, is how the local population can be included in the decision making process. This method of reaching political decisions is unfortunately still in its infancy in China. The concept of biosphere reserves in terms of participative approaches can provide new ideas and practical experiences in this respect. The issues of land ownership and usufruct are also breaking new ground. In principle, there is only long-term usufruct and not private ownership of land in China. For example, the management of the Yancheng BR is made more difficult by the fact that the administration is responsible for the use of the land in the core area of the biosphere reserve, but other administrations or private individuals own the rights to the use of the buffer zone and transition area. The spread of public and private property ownership that is quite normal in Germany, leads to considerable problems in China at a local level. This is because suitable, participative approaches to solving conflicts are still not used. Another significant problem in the Chinese biosphere reserves is (mass) eco-tourism. In the Huanglong BR the annual number of visitors rose from 10,000 in 1983 to 800,000 in 2001 and the numbers are still rising (Chinese MAB National Committee 2003). The German visitors were able to find out about impressive, successful methods of managing visitor flows here.
In the Juizhaigou BR, visited at the same time, the upper limit of daily visitors, for example, has been restricted to 10,000. However, it remains unclear as to whether this number actually represents the maximum limit for the area. Research is needed to clarify this issue. Incidentally, the area is closed to private cars and for the entrance price of approximately € 30,

Managing visitor flows in the Huanglong Biosphere Reserve

visitors can travel around freely in buses fuelled by natural gas. The Biosphere Reserve provides work for around 1,000 people, including rangers, drivers, waiters or souvenir sellers. They earn more through these activities than in agriculture, which is extremely arduous and unproductive on the steep mountain slopes. For this reason, many of the inhabitants give up terrace farming. Not only does this result in an increase in soil erosion and a threat of landslides onto roads, but also valuable, traditional cultural landscapes disappear. This example clearly shows that biosphere reserves can only achieve the required results of sustainable regional development if all ecological, economic and socio-cultural aspects are considered.

"The lucky fairy" in the biosphere reserve

Translated, the Tibetan name for the well-meaning guide in the Jiuzhaigou Biosphere Reserve means, "the lucky fairy". The enthusiasm and joy she shows as she guides the visitors through the Biosphere Reserve is so infectious that the rainy weather is quickly forgotten during the walk, which lasts several hours along the numerous lakes and waterfalls. In addition to providing many interesting details on the nature and culture of the region, she talks about her education at an education college and her Tibetan family with her six brothers, who manage a small farm at the edge of the Biosphere Reserve. Thanks to "the lucky fairy", the Juizhaigou Biosphere Reserve remains not only a spectacular natural beauty but also a biosphere and natural habitat and home to friendly people that will not be forgotten.

(Text and photograph: Getrud Hein)

The Huanglong und Juizhaigou Biosphere Reserves are exceptionally successful economically. With admission charges of approximately € 15 or € 30 they are practically "money-making machines", which make a considerable contribution to the regional economy. It is almost impossible to improve the management of the large streams, of people which would overwhelm Central Europeans. However, a pure conservationist certainly needs to get used to the image of crowds of people amongst nature worthy of protection.

Thirdly, during a visit to the Wolong Biosphere Reserve, lasting unfortunately only several hours, the German visitors were introduced to an example of the activities China is undertaking to protect endangered species. There is a breeding station for about 50 pandas in the Biosphere Reserve. According to the director of the station, in 2002 six young pandas were born, which is a real achievement if the great difficulties that exist with breeding pandas in captivity are taken into consideration. When the mother has finished weaning, most of the animals are given to Chinese zoos. Pandas are threatened with extinction and are in Annex I of CITES.

Young pandas in the breeding station in Wolong

Pandas are considered almost sacred by the Chinese: in 2003 a poacher was sentenced to 15 years in prison.

In general, the bilateral cooperation with the Chinese MAB National Committee is an example of how important it is to be open-minded and learn from others. By putting individual problems and successes in an international context, they can be seen in relative terms.

Particularly for an international programme such as the MAB Programme, this represents both a basic requirement and an objective to be met. Despite the differences in both countries we have seen a lot of common ground in the problems associated with a sustainable regional development. We also share the belief that lasting concepts for the future can only be delivered by taking economic, ecological and socio-cultural components into consideration at the same time. Our belief that UNESCO biosphere reserves are suitable models for demonstrating methods to achieve sustainable development has also been confirmed.

The cooperation is to be continued in the coming years. The question in everyone's interest is: How can the economic components of the biosphere reserves be improved? There are plans to hold a bilateral workshop in Berlin on the subject of "Quality Economies in Biosphere Reserves" and UNESCO representatives will also take part in this.

In addition to this, at the end of August 2004 a bilateral conference on the subject of "Biosphere Reserves as an Approach for Sustainable, Environmentally-Friendly Use of Ecosystems within the Temperate Zone" will take place at the German and Chinese Science Centre in Beijing.

Literature

UNESCO (1996): Final Report of the Co-operative Ecological Research Project (CERP), Paris.

Chinese MAB National Committee (2003): Newsletter Number 15.

4.14 *Transboundary Biosphere Reserves: Win-Win Solutions for People and Nature*

Elke Steinmetz

The Idea

Borders that have developed politically and historically very rarely concur with "ecological borders". Ecosystems therefore often cross political borders. These can separate both nation states and administrative regions (*Länder*), e.g. in a federal system such as that of the Federal Republic of Germany.

Although species of fauna and flora often find relatively unhampered means of dispersal, their joint protection and sustainable use is often difficult due to the different legislative, administrative and political conditions in the states or *Länder* concerned. Geographically integrated ecosystem and conservation area management is therefore only possible if political borders are permeable.

However, cross-border nature conservation projects are not only important for the protection and conservation of biodiversity, but also for its sustainable use. Cooperation across borders in nature conservation can also play an important role for international understanding, trust building and for dealing with ethno-political tensions and conflicts. This is how so-called win-win situations can be achieved both for people and for people and nature.

Win-win solutions are solutions that solve conflicts on the basis of shared interests or a balance of interests. The term became famous under the so-called Harvard Concept (Fischer R. et al. 1993 and Ropers, N. 1995). The key is that there is a gain or a benefit for each of the conflicting partners (win-win) that is based on non-competing individual interests or on shared interests. In the example of transboundary biosphere reserves, where solutions for nature and people and for people and people can be found, an ideal "win-win-win" solution can therefore be achieved.

The idea of combining the protection and use of natural resources and the mitigation of social conflicts is not new: as early as 1932 the "Waterton Glacier International Peace Park" was established between the USA and Canada to underline the long-lasting peace between the two countries (Sandwith, T. et al. 2001). In 1935 another National Park was set up between the USA and Mexico ("Maderas del Carmen and Canyon de Santa Elena/Big Bend National Park"), which was to help to solve border problems.

In recent years, the establishment of transboundary "parks for peace" has been advanced and supported by various international organisations (IUCN, WWF, CI and PPF), especially in southern Africa (World Conservation Union 1997). Besides this conservation category of national parks, transboundary biosphere reserves with their integrated, holistic concept are a particularly suitable instrument for generating positive synergetic effects from cooperation.

The Concept

Within the World Network of Biosphere Reserves the numbers of transboundary biosphere reserves are rising. These are biosphere reserves that are on both (or all) sides of a political border. Recognition as a transboundary biosphere reserve by UNESCO means that they are officially announcing their intention to cooperate in the protection and sustainable use of an ecosystem by means of joint management (UNESCO 2000).

Furthermore, recognition includes an agreement to implement the "Seville Strategy for Biosphere Reserves". Within the context of this strategy adopted in 1995 at an international conference of experts in Seville (UNESCO 1996), steps for the further development of the biosphere reserves in the 21st century were recommended. One of the goals of the strategy is to "encourage the establishment of transboundary biosphere reserves as a means of dealing with the conservation of organisms, ecosystems, and genetic resources that cross national boundaries" (UNESCO 1996).

The follow-up conference "Seville + 5" in Pamplona in the year 2000 adopted special recommendations for the establishment of transboundary biosphere reserves in order to emphasise their particular importance (UNESCO 2000). These "Pamplona Recommendations" describe the process to set up a transboundary biosphere reserve, its function, its general institutional conditions and the link to the objectives of the Seville Strategy. The idea of combining joint management structures for the protection and sustainable use of biodiversity with mechanisms and structures for cooperation and conflict prevention is a common theme through the Recommendations.

The management of transboundary biosphere reserves has to cope with many challenges and obstacles, especially in administrative, legislative and financial matters. This can be a disadvantage in the effectiveness of the management. By contrast, however, there are the following main advantages (EUROPARC Federation 2001):

- successful conservation and sustainable use management;
- promotion of regional development in peripheral areas;
- preservation of cultural identity and integrity as well as
- successful mitigation of conflicts, promotion of peace and crisis prevention.

Added value which can develop transboundary biosphere reserves into an instrument of sustainable environmental, developmental and security policy for the 21st century can be achieved by combining these advantages and, in particular, by integrating the last aspect.

So far, UNESCO has officially recognised six transboundary biosphere reserves (as of May 2003):

- Region "W" (Benin/Burkina Faso/Niger)
- Krkokonose/Karkonosze (Czech Republic/Poland)
- Vosges du Nord/Palatinate Forest (France/Germany)
- Eastern Carpathians (Poland/Slovakia/Ukraine)
- Tatra (Poland/Slovakia)
- Danube Delta (Romania/Ukraine)

With the exception of the Region "W", all of the recognised transboundary biosphere reserves are in Europe. But world-wide countless efforts are being made to get existing and developing cooperation between biosphere reserves recognised by UNESCO and, thus, institutionalised. For example, these include the demilitarised zone between the two Koreas, the Panama/Costa Rica and the Bolivia/Peru border regions (BRIDGEWATER, P. 2002) as well as Altai (Russia, China, Mongolia, Kazakhstan).

Three examples of transboundary biosphere reserves

- Rhön Biosphere Reserve in three *Länder* (Bavaria/Hesse/Thuringia):

The Rhön Biosphere Reserve lies in the centre of Germany in the borderland between the three Federal *Länder* of Bavaria, Hesse and Thuringia and is thus a national transboundary biosphere reserve within the federal system of Germany. Due to the political, administrative and legislative differences in the three *Länder*, the Rhön Biosphere Reserve is comparable to a biosphere reserve that crosses the borders of international sovereign states, even if it cannot officially be listed as such in the UNESCO nomenclature.

The "Upper Rhön" at the heart of the Biosphere Reserve has given the Rhön the name "Land of open vistas".

As a typical medium-range mountain landscape, the Rhön has a landscape that has come about from extensive agricultural use, which has given the Rhön the name "Land of open vistas" due to its open landscapes. The conservation and sustainable use of this landscape largely determines the shared guiding principle of the Rhön Biosphere Reserve.

- Palatinate Forest-North Vosges transboundary Biosphere Reserve (Germany/France):

The Palatinate Forest-North Vosges transboundary Biosphere Reserve lies in the German-French border region and has been officially recognised by UNESCO as a transboundary biosphere reserve in 1998. It was formed from the North Vosges Regional Nature Park and the Palatinate Forest Nature Park. In many respects, the area of the Palatinate Forest-North Vosges transboundary Biosphere Reserve is one unit, both from an ecological and landscape point of view and on the basis of its shared cultural and historical development. Key aspects of the work in the Biosphere Reserve are the conservation and sustainable use of the largest contiguous forest area in western Europe, environmental education, the economic development of the region, intercultural approaches and international cooperation.

The cross-border discovery route "Landscape across Borders" is a good example for the implementation of cross-border environmental education measures: four circular walking routes link up Alsace, the Palatinate and Lorraine and are an invitation to discover the German-French model region.

The German-French expert working group „Biodiversity and Nature Conservation"

- **Altai transboundary Biosphere Reserve:**
Currently (June 2003), efforts are being made to establish a quadrilateral, transboundary biosphere reserve in the Altai region. The Altai mountains lie in the border region between the Russian Federation, Mongolia, China and Kazakhstan. Geographically, they form the border between the Siberian taiga and the steppes and deserts of central Asia. According to the Worldwide Fund for Nature (WWF), Altai is one of the 200 regions on earth with the greatest species diversity, in other words it is a "hotspot of biodiversity". However, as well as the diversity of habitats, Altai also has a special cultural diversity: the population is made up of Altaians, Kazakhs, Mongols, Uigurs, and Tuvans as well as Russians, Germans and Chinese. A feasibility study or studies coordinated by the *Deutsche Gesellschaft für Technische Zusammenarbeit* (German Technical Cooperation, GTZ) is/are currently (2003) examining whether a transboundary biosphere reserve is the appropriate instrument for achieving the goals of nature and landscape conservation in this region as well as those of sustainable socio-economic regional development. The results of this study should be available in December 2003.

The exchange of information and experience plays an important role in establishing a transboundary biosphere reserve. In this way, shared traditions and customs can build a bridge, as can be seen here in the welcoming ceremony for a group of visitors in the Russian part of Altai.

What Does the Future Hold?

Due to added value, transboundary biosphere reserves offer great potential for sustainable development. However, to make this added value transferable and usable, there is a great need for process and action oriented research as well as the participation of all stakeholders involved.

Whereas many research projects – ongoing and completed – are already studying nature conservation aspects of cross-border conservation area management in biosphere reserves, there are only a few research projects with a socio-economic or socio-cultural focus dealing with the often very important "soft factors". For this reason, at the World Parks Congress to be held in Durban (South Africa) in September 2003 the "International Working Group on Transboundary Protected Areas" of the World Conservation Union will call for this deficit in research to be remedied. Some of the questions that will be important in the future are:

- How can the added value of transboundary biosphere reserves be made usable for the particular region from an ecological, economic, socio-cultural and political point of view?
- What general conditions and criteria characterise transboundary biosphere reserves to achieve synergetic effects in the success of nature conservation projects and conflict mitigation and prevention? What role can they play in development cooperation?
- What formal and informal structures are suitable for a particular transboundary biosphere reserve to make the added value usable?
- What mechanisms are appropriate to use cultural differences positively and as a dynamic impetus for development? What role do intercultural communication patterns play here?
- Under what conditions can a "transformation" from destructive to constructive conflicts be achieved by transboundary biosphere reserves (Ropers, N. 1999)?

These questions show just a very small excerpt, but give an impression of the "wide area" between sociology, political science, organisational research, psychology, communications science, peace and conflict research as well as ethnology that is touched on here. They are relevant both to biosphere reserves in Germany and other industrialised countries and to biosphere reserve in developing countries (cf GTZ 2000). What is important for the questions cited above and those that have not been cited is to examine case studies and to derive transferable, documented and evaluated lessons learned and best practices based on monitoring, analyses and evaluation. The UNESCO World Network of Biosphere Reserves offers ideal conditions for this and should be used for an exchange of experience and for applied research.

Biosphere reserves are model regions. Here it should be possible to observe and evaluate successful processes and to find practicable and transferable solutions for sustainable regional development that gives equal consideration to people and nature. Furthermore, transboundary biosphere reserves are international model regions that are a particular challenge and opportunity for transdisciplinary and interdisciplinary research, testing and development of sustainable regional development strategies.

Literature

BRIDGEWATER, P. (2002): Grenzüberschreitende Biosphärenreservate. Vergangenheit, Gegenwart und definitiv die Zukunft! In: BIOSPHÄRENRESERVAT NATURPARK PFÄLZERWALD (Ed.) (2002): Zwischen Frankreich und Deutschland hat die Natur keine Grenze mehr. Dokumentation.

EUROPARC FEDERATION (Ed.) (2000): Basic Standards for Transfrontier Cooperation between European Protected Areas.

EUROPARC FEDERATION (Ed.) (2001): EUROPARC Expertise Exchange Working Group on Transfrontier Protected Areas.

FISHER, R., URY, W. & B. PATTON (1993): Das Harvard-Konzept: Sachgerecht verhandeln - erfolgreich verhandeln.

GTZ (Deutsche Gesellschaft für technische Zusammenarbeit) (Ed.) (2000): Krisenprävention und Konfliktbearbeitung in der Technischen Zusammenarbeit.

ROPERS, N. (1995): Friedliche Einmischung. Strukturen, Prozesse und Strategien zur konstruktiven Bearbeitung ethnopolitischer Konflikte. In: Berghof Forschungszentrum für konstruktive Konfliktbearbeitung (1995). Berghof Report Nr. 1.

ROPERS, N. (1999): zur Begrifflichkeit im Arbeitsfeld „Konfliktbearbeitung" und „Friedensförderung". Unpublished Manuscript.

SANDWITH, T. et al. (2001): Transboundary Protected Areas for Peace and Co-operation. IUCN.

UNESCO (Ed.) (1996): Biosphere Reserves. The Seville Strategy and the Statutory Framework of the World Network, Paris.

UNESCO (2000): MAB Sevilla + 5 Recommendations for the Establishment and Functioning of Transboundary Biosphere Reserves.
http://www.unesco.org/mab/mabicc/2000/eng/TBREng.htm (Access June 2003)

WORLD CONSERVATION UNION (IUCN) (1997): Parks for Peace. – Parks. The International Journal for Protected Area Managers, Vol 7, No 3.

Acknowledgements

This work was supported by the individual project IP7 "Environmental Change and Conflict Transformation" within the context of the research focus NCCR North-South "Research Partnerships for Mitigating Syndromes of Global Change", financed by the Swiss National Fund to Promote Scientific Research (SNSF) and the Directorate for Development and Cooperation (DEZA).

Examples from Research

5.

5.1 *Research and Monitoring in German Biosphere Reserves: An Overview*

Birgit Heinze

The UNESCO programme Man and the Biosphere (MAB) is one of several scientific research programmes run by UNESCO, but it is the only one that, with its biosphere reserves, has uniformly defined areas to research the relationship between humans and nature. The Worldwide Network currently (2003) comprises 440 model regions in 97 countries. The areas of UNESCO biosphere reserves are divided into core areas, buffer zones and transition areas, each of the zones being assigned to specific functions with graduated human impact. This means that all components of sustainable development can be monitored, researched and tested in the long term in biosphere reserves.

Research and monitoring are major goals of the Programme and a fundamental task for all biosphere reserves.

Since the UN Conference on Environment and Development in Rio de Janeiro in 1992 the international community has committed itself to sustainable development. However, interdisciplinary research into sustainability has problems in financing itself because it is caught between all stools of the sectoral structure of our research scene (and also that of other countries). It cannot be assigned to economic, ecological or social science research. For this reason, from 2004 the Federal Ministry for Education and Science (BMBF) will be replacing the Federal Government's Environmental Research Programme with two new general programmes: one of them is to draw up implementation-oriented contributions towards realising sustainable development and the other is entitled "Vulnerability of the System Earth". At European level, too, there are corresponding moves in the research funding scene that will at last make research in sustainable development possible. As soon as sustainability research is funded by the donor institutions, the Network of UNESCO Biosphere Reserves provides representative landscape areas for research all over the world that are especially suited to the realisation of projects. Thus, in future, biosphere reserves will be able to gain in importance as reference, research and testing grounds for sustainable development.

The following is an overview of the research activities in the German biosphere reserves.

All of the information is based on a questionnaire from the German MAB-Secretariat of June 2003. The feedback from 19 administrative offices has been evaluated, representing 13 German biosphere reserves. The administration from one biosphere reserve did not feel able to provide information, which unfortunately means that this overview cannot give a complete picture of the biosphere reserves in Germany.

Frame Research Plan

A frame research plan has been set up in only three of the 19 administrations, such a plan is scheduled in two more. Existing frame research plans are usually of a non-binding nature because they depend on budgetary and staff resources. Otherwise, the chapter on research of the framework concepts that are obligatory for biosphere reserves is used to compile the relevant research questions for each biosphere reserve.

Research Projects

This question dealt with the number of projects and their assignment to main thematic issues. The questionnaire also asked about the financial volume in each case; however, the feedback in this respect is so sparse that it does not make sense to evaluate it. The number of projects allows only a limited conclusion about the actual scope of research because big and small research projects with long or short running periods are treated equally. In spite of these restrictions, the compilation of the information gives a first impression:

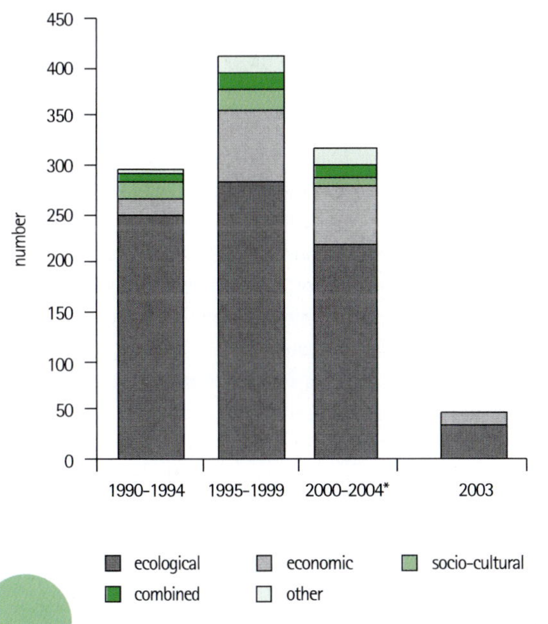

Fig. 1: Number of research projects, summarised in five-year intervals, and with thematic assignment.
[: this also included those projects that will probably run in 2004. The bar for 2003 reflects the current annual status.]*

The number of research projects changed in the three five-year intervals portrayed, from 295 to 411 to 317; 49 projects were indicated for 2003.

Fig. 2 reflects the thematic assignment of the research projects: the main research was and is in ecological subjects, a rising trend of projects in the economic subjects can be seen; the number of projects with a socio-cultural emphasis is low.

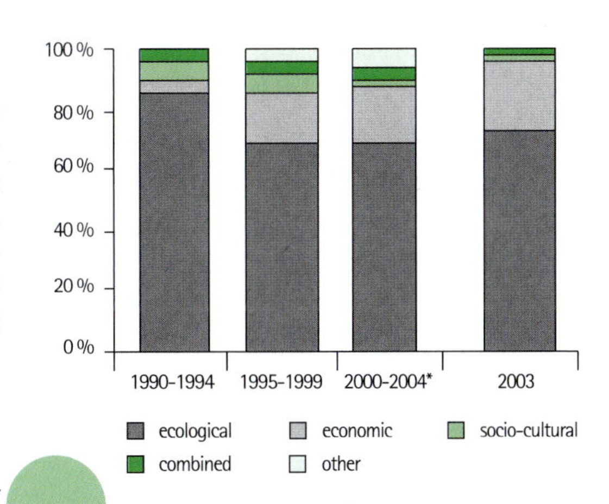

Fig. 2: Thematic assignment of the research projects
[: this also included those projects that will probably run in 2004. The bar for 2003 reflects the current annual status.]*

Transdisciplinary and interdisciplinary projects account for a shrinking proportion, in other words those projects that represent "original" sustainability research. But since the list is based only on the number of projects, this result should be interpreted with caution: an interdisciplinary project on sustainability will have a larger financial volume and a longer term. The departmental dependency of finances for research projects is also reflected in these results.

Undergraduate and Doctoral Theses in Biosphere Reserves

Numerous research issues in biosphere reserves are dealt with in undergraduate and doctoral theses. Figs. 3 and 4 show, once again in five-year intervals, the number and the thematic assignment of the theses.

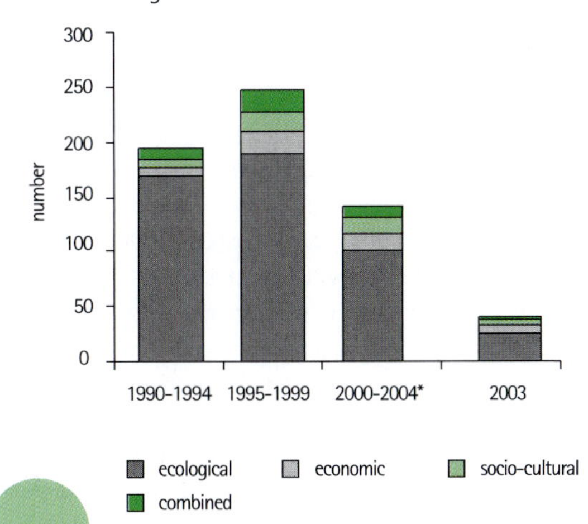

Fig. 3: Undergraduate theses in German biosphere reserves
[: this also included those projects that will probably run in 2004. The bar for 2003 reflects the current annual status.]*

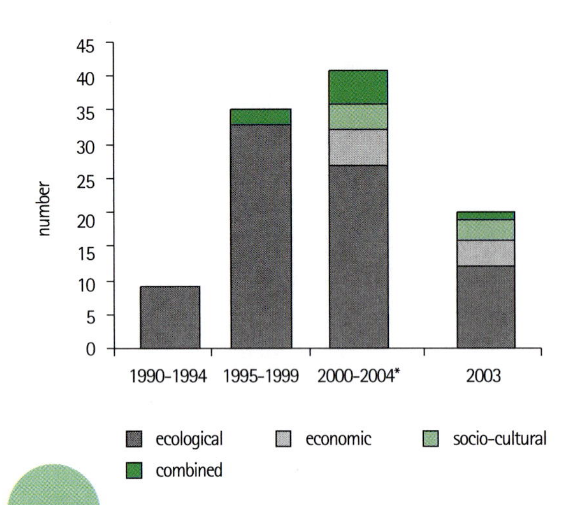

Fig. 4: Doctoral theses in German biosphere reserves
[: this also included those projects that will probably run in 2004. The bar for 2003 reflects the current annual status.]*

The low number of undergraduate theses in the last five-year interval, contrasted with an increased number of doctoral theses in the same period, is noticeable. Biosphere reserves are obviously suitable for more complex studies. As far as the subject matter is concerned, the majority – as in the research projects – are in the ecological subjects. The increase in combined, i.e. transdisciplinary and interdisciplinary theses, in doctoral theses is pleasing.
The questionnaire also asked about the involvement of foreign scientists. However, foreign undergraduates and PhD students were only listed here and there. Those named came from the Netherlands, Spain, Denmark, Ethiopia and the USA, among other countries.

Cooperation with National Institutions

A ranking of cooperation with national institutions (for the last five years) reveals that the 19 BR administrations mainly cooperate with universities, specialist authorities from the *Länder* and with universities of applied sciences. The Federal Ministry for the Environment, Nature Conservation and Nuclear Safety (BMU) with its subordinate specialist agencies Federal Environmental Agency (UBA) and Federal Agency for Nature Conservation (BfN) is cited as the next important cooperation partner. The Federal Ministry for Education and Research (BMBF) is also cited often, but with lower priority. Other Federal specialist authorities and further institutions are mentioned lower down in the ranking and only here and there. What is remarkable is that only two biosphere reserves cite cooperation with the Federal Ministry for Consumer Protection, Food and Agriculture (BMVEL), and that with very low priority.

Cooperation with International Institutions

Cooperation with the EU Commission, UNESCO and its scientific programmes and with foreign universities and research institutions dominates here (with regard to the last five years). It is surprising that some biosphere reserve administrations apparently have no international contacts or cooperation in research at all.

Knowledge of Research Work in the Biosphere Reserve, Safeguarding Results

One question in the questionnaire was concerned with whether and how it is ensured that the results of scientific work conducted in a biosphere reserve are brought to the attention of its administration.
The administrations of the biosphere reserves have secured knowledge of all research work that need a permit or an exemption or are subject to an obligation to report; this mainly concerns ecological projects in the core areas and buffer zones. The research work that they commission themselves or that they co-finance are thoroughly recorded by the BR administrations. In these cases, the hand-over of the results is ensured as a condition or via agreements.
Moreover, researchers, groups of researchers, scientific institutions and authorities are happy to use the support and cooperation of the BR administrations, be it because of the local and personal knowledge there or because of the data available there. In these cases, too, the working results are usually handed over to the BR administration.
One BR administration states that it always knows about all of the research work because of the small, manageable area of the biosphere reserve; another administration admitted that it had no knowledge of the research work ongoing in the area.
This shows that there is no generally binding mechanism to safeguard the results of research work in the biosphere reserve, especially with regard to non-ecological issues in the buffer zones and transition areas. But good contacts to the institutions conducting research here appear to compensate for this.

Communication of the Research Results

The results of the research works are communicated (in the order of frequency) in the public media, in specialist publications, in lectures and in the internet. Around half of the BR

administrations maintain their own database on research projects and almost all archive the results.
For the international community the results are shown in specialist publications, but mainly in lectures to foreign visitors. A special presentation in the internet is the exception.

Implementation of the Results

Here, the questionnaire asked about the estimated percentage of implementation of the results "in your own biosphere reserve", "in other biosphere reserves in Germany", "in Germany outside biosphere reserves", "in other biosphere reserves worldwide" and "no implementation". Only sporadic information could be evaluated. The assessment is apparently difficult. The interpretation that the questions were of no significance to most BR administrations also appears appropriate.

Monitoring in Biosphere Reserves

Reports of socio-economic parameters were the exception with respect to the question of monitoring in biosphere reserves. Here, the activities of the BR administrations are currently limited to documenting visitor numbers to the biosphere reserve or the information centres and participant numbers in specialist guided tours. This means that the activities of individual biosphere reserves are all the more remarkable, such as traffic censuses or recording large quarries to develop a conservation concept in the South-East Rügen BR, economic monitoring in the Spree Forest BR, where a set of approximately 42 parameters (including the natural yields of all crops, energy inputs and emissions, animal stocks) are documented in ten reference farms or the monitoring of farms and fisheries (animal stocking, land development) and the skilled crafts companies in the Oberlausitz Heath and Pond Landscape BR.
The focus of monitoring is currently on environmental monitoring in biosphere reserves in Germany. This is described in more detail in Chapter 5.3.
A large amount of data are not collected separately by the BR administrations, but taken over from other, existing data collection and are compiled in line with needs (e.g. population and unemployment figures).
The data collections and measurements in the biosphere reserves are conducted by administration staff (e.g. nature rangers), volunteers (e.g. in bird or butterfly censuses) or by scientists or institutes commissioned for this purpose.
The costs of a monitoring programme therefore vary accordingly. They range from € 0.00 when volunteers and students are deployed up to € 400,000.00 per year for ecosystemic environmental monitoring.
The results are used for status reports and also predictions for future developments and trends to test the impact of the measures implemented or also to decide whether a measure should be initiated or not.

Use of Geographic Information Systems (GIS)

Geographic Information Systems (GIS) are playing an increasingly important role in monitoring. Here, the data and measured results are assigned to the survey locations and

Monitoring

Monitoring is important for understanding changes. This is one of the main tasks of biosphere reserves in their function as reference areas. In monitoring, certain measuring variables, so-called parameters (e.g. temperature) are measured and continuously documented at regular intervals. Depending on the question and financial and staff scope, the measurements are taken at intervals of days, weeks, months, years or several years or even decades. Monitoring can affect animal and plant species, biotope types and ecosystems just as much as soil and water quality or climate factors, the groundwater level, snowslip areas, vegetation borders, hoofed game grazing, the services in contractual nature conservation as well as unemployment figures, age and sex of the population, tax revenues, number of companies, jobs, etc. The aim of this monitoring is at any time to be able to state the status and the change of a parameter over time, compare it to comparable values and make predictions. In the ecological area in particular the natural dynamism of ecosystems is monitored so that it can be assessed at all.
Since measuring each individual parameter is not possible, researchers in the ecological area in particular often use so-called indicators, whose changes are representative of broader themes and allow broader statements (e.g. frequency of earthworms as an indicator of soil quality).

saved in big databases. This means that they can be called up at any time and can be compared and combined with data from other databases specific to the location.
Almost all biosphere reserves in Germany state that they have set up Geographic Information Systems; only in a few cases do they use the corresponding system from another place. The systems are continuously expanded and are used in many ways, e.g. for plans or to compile card material.

Human Resources

To put the work in the field of research and monitoring in proportion to the BR administration personnel available for it, the biosphere reserves were asked about the number of employees with a scientific training. In five administrative offices there are no scientific employees whatsoever, three cite one or two scientists, three administrative offices five to seven and two ten. But they conduct scientific work only in exceptional cases. They are mainly given administrative tasks, perform sovereign duties or are involved in project management.
There is very little specialist knowledge from economic or social science fields. Only one biosphere reserve cited a graduate educationalist as an employee.

Financial Resources

Of the 19 administrative offices, only three have their own research budget, ranging from € 12,000 to € 300,000 per year. One biosphere reserve was given the prospect of its own budget for 2003.
The BR administrations finance their research work with third-party funding. Here, a ranking showed that the relevant Federal *Länder* are in first place, the Federal Government in second, the EU in third and universities in fourth place. Other providers of third-party funding of much less importance were local authorities, industry/commerce and sponsoring associations.

Proposals

With these rather scarce staffing and financial resources in the BR administrations it is hardly surprising that appropriate (minimum) staffing, financial and technical resources are one of the most cited proposals for improvement in the field of research and monitoring, which the questionnaire also asked about. Furthermore, the following were also cited:
- own research budget;
- improvement in means of communication between the BR administration and potential communicators of information;
- central database for research projects for better coordination between all German biosphere reserves;
- updating the harmonisation of main research interests in the various biosphere reserves in Germany;
- concept for research and environmental monitoring for the individual biosphere reserves;
- general research plan for every biosphere reserve in Germany, drawn up jointly with the local players;
- promoting and publicising biosphere reserves as research areas;
- uniform, binding framework guideline;
- uniform database and communication infrastructure;
- incorporation of all biosphere reserves in monitoring all over Germany;
- drawing up an inventory of ongoing research programmes in the German biosphere reserves;
- frame research plan for all biosphere reserves taking account of international requirements.

This list illustrates that there is great interest in research – and that there is a great need on the part of the BR administrations. The following contributions will present a few research projects as examples, which nevertheless demonstrate that fundamental and innovative, theoretical and application-related issues are already being intensively researched.

Acknowledgements

At this point, I would like to thank all BR administrations for responding to the questionnaire, which was, after all, only one of so many!

5.2 Regional Marketing of Agricultural Produce in German Biosphere Reserves

Armin Kullmann

The R&D Project "Regional Marketing in Biosphere Reserves"

The function of biosphere reserves, originally focused more on nature conservation, environmental research and education, was further developed at the MAB Conference in Seville in 1995. Ever since, the intention has been for the biosphere reserves also to act as model regions for sustainable development (UNESCO 1996). A core function of the biosphere reserves has been to "green" land use. At the MAB Conference in

Factors for the Success of Project Management	Brief Explanations
Motivation of the regional participants	*Pressure of problem; Problem awareness; Willingness to change; Commitment; Pursuit of profit; Use of personal capital*
Committed key people	*Motivators; People fully committed to the project, who take action, behave as leaders, initiate the project, motivate colleagues*
Process expertise	*Ability to lead a group, to manage a project, to develop an organisation, to implement the factors for success*
Influential partners	*Acquisition of socially, politically or economically influential partners such as associations, churches, companies etc.; Conservation areas, conservationists and farmers are influential partners*
Good relationships	*Good personal contacts to decision-makers (administrative district heads, office leaders, ministers), to responsible specialists and to the public; Proactive issue management*
Access to resources	*Availability of working hours and financial means through support programmes, political or public support*
Organisational structure	*1. Project management: clear objectives, tools, processes 2. Organisational development: legal basis, personnel, business management (of the project and the business)*
Win-win situation	*Cooperation instead of conflict with different interest groups; all must achieve profit or benefits*
Demonstrable results	*Achieve results, including economic success; Communicate successful results; Produce a consistent series of results; Long-term development is more important than short-term success*
Factors for the Success of Marketing Management	
Marketing expertise	*Market knowledge and market contacts; Experience in production, processing, sales, business management, personnel and business leadership; if necessary, qualification or external consultancy*
Consistent marketing strategies	*General marketing principles: Unique Selling Point to enable competitive advantage, understanding the marketing tool-kit*
Appropriate geographic delimitation	*Region of a certain size (e.g. an administrative district) for sufficient quantitative and qualitative supply; endogenous demand dependent on number of inhabitants; regional identity of the area is important*
Definition of specific production guidelines	*Regional brands mostly indicate region of production and quality; a regional focus alone will not provide sufficient competitive advantage; Protection of animals, the well-being of all species, unspoilt nature and transparency are more important*
Effective controlling system	*Promises to the customer must be kept; Guidelines should always be controlled effectively; Independent checkpoints and control bodies; Principles of crisis management*
Top quality products and services	*The most important success factor; Taste, smell, sight, touch, texture etc. are deciding factors; Packaging is important for the image; Customer-orientated services are increasingly important*
Acceptable price-benefit relationship	*Dependent on the price policy in the marketing strategy: average or up-market segment? Target groups? End customers or retailers? Quality and image adapted accordingly*
Problem-free distribution	*Ability to target and achieve (markets, sales methods and organisation); Clear identity (features, brands); Technical requirements (storage, cooling methods, vehicles etc.); Reliable, flexible logistics*
Professional communication	*Corporate design (logo, brand); Advertising focused on the target group; Sales support at the point of sale; Press and public relations work*

Table: Further-developed factors for success in regional marketing, with brief explanations

Pamplona in 2000 the development of sustainable economics was put at the very top of the agenda for the biosphere reserves by setting up a "Task Force on Quality Economies".
In this respect, the marketing of agricultural, forestry, fishing and wine products etc., produced on a sustainable basis, is particularly effective, as it combines aspects of conservation with economic benefits for the land users. The goal of regional marketing is, above all, to support local material and economic cycles (regional closed substance cycles) in line with the well-known motto "from the region – for the region".
Within the framework of the research and development project (R&D project) "Environmentally-friendly regional development through product and regional marketing with the example of biosphere reserves" a study was carried out to see if German biosphere reserves, compared to other regions, actually do have a model function with regard to local food marketing.
In this connection, factors for the success of regional marketing projects were identified and tested in an evaluation of ten selected model regional marketing projects outside the biosphere reserves. On the basis of this evaluation, the first set of factors for success was further developed and supplemented by key marketing factors.
The table (cf p. 119) includes these additional factors for success, which have been used for the status quo analysis of regional marketing in biosphere reserves (Kullmann, A. 2003 a). On this basis, at the end of the R&D Project a practical method was developed for the analysis of the factors for success (Kullmann, A. 2003 c).

Selected Results from the Status Quo Analysis

Satisfaction with implementation of the factors for success

Representatives from the 20 administrative authorities of the 14 German biosphere reserves were questioned. The factors for success tested were regarded more or less as highly by these experts as by the experts of the model projects previously.
It was evident, however, on assessing the resulting levels of satisfaction of the experts that there was a greater feeling of discontent among the biosphere reserves than among the model projects. The factors causing most dissatisfaction were the resources, organisational structure, communication, influential partners and key people involved (Kullmann, A. 2003 b).
Only a few biosphere reserves are model regions for regional marketing. There are considerable differences in the type, number, scale and degrees of success of the regional marketing activities and projects in the German biosphere reserves.

The Rhön Biosphere Reserve can be considered a leading model region nationwide for regional marketing focusing on sustainability (cf Chapter 4.1).
Despite their differences, the Schaalsee, Schorfheide-Chorin and Spree Forest Biosphere Reserves can also be considered as models. At the time of the study, they had already introduced regional brands. In recent years, extensive and successful regional activities have been implemented in four other biosphere reserves. Eight biosphere reserves have demonstrated minimal or no activity to date.

Suitability as production regions

More than half of the biosphere reserves do not have a transition area with land or water that can be used for producing food (e.g. national parks) or they are too small or too limited structurally to form a significant production region. They would have to identify transition areas or cooperate with businesses outside the biosphere reserves. Two of the three biosphere reserves with regional brands have therefore extended their territory clearly beyond the biosphere reserve itself. In principle this seems sensible in order to include external players.
With the goal of a sustainable economy in mind, in the future biosphere reserves should probably be delimited geographically both according to nature and landscapes as well as socio-economic criteria. Up to then, adjacent administrative districts could be included in order to form a sufficiently large production region.

Production and quality guidelines

Only the three biosphere reserves with regional brands and the Rhön BR had drawn up criteria for production in their region by mid-2003. Defining the guidelines is usually the subject of long and sometimes very intense discussions. They can only be determined by a normative and strategic marketing decision of general principle. The standards for organic agriculture only underpin approximately 35 per cent of all German regional projects (www.reginet.de). In addition, many BR administrations also feel committed to conventional farming, which should be supported in its goal to achieve greater sustainability.
The definition of the production criteria and the brand focus vary significantly among the biosphere reserves. Whereas in the Rhön, it is predominantly the marketing of organic products which has been supported to date, the regional branding of the Spree Forest, on the other hand, stands for purely conventional agriculture.
Product quality is considered by all experts to be the most important factor for the success of regional marketing. This should lead to initiatives that will incorporate quality management into regional marketing, e.g. by using annual awards.

Participants, products, marketing methods

Direct marketing is the basis of regional marketing. Butchers, bakers, the processing trade and industry, the retail and catering trades and accommodation businesses are equally important participants. The aim should always be for these to become participants and partners. As in all types of marketing, there is great diversity in the size, forms of organisation and marketing strategies of regional marketing projects.

The KFF meat processing plant of Tegut, the leading organic food retailer through its own supermarkets in Europe, is a very important customer for organic free-range beef from the "BR Rhön partner farms".

The marketing strategies for tourist regions and areas less developed for tourism are fundamentally different. For the biosphere reserves that have been developed for tourists, holiday-makers and day trippers represent an important customer group. Clearly defined strategies and a higher level of quality should form the basis of tourism and product marketing. There should be high quality marketing and public relations of the biosphere reserves in all respects.

There is an important difference between West and East Germany, i. e. the former Federal Republic of Germany and the former German Democratic Republic. The agricultural structure in the new Federal *Länder* consists predominantly of large farms, which rarely process their own produce or market it regionally. Similar to the situation in West Germany, there is a need here to rebuild the working structures cooperatively or independently. This requires political support.

Regional marketing cannot be limited to the region of production alone in any region. Neighbouring towns are important markets. A future strategy of regional marketing could be to supply independent food retailers.

Requirements for qualification

The demanding qualification requirements for the regional participants in local marketing have been rather unexpected. BR managers and stakeholders have attached great importance to acquiring and "raising influential parties" and to qualifying specialist and leadership staff in businesses.

At the same time, the "Biosphere Job Motor" has been developed in the South East Rügen and Schaalsee Biosphere Reserves (cf Chapter 4.3). In the Rhön Biosphere Reserve, a project for the qualification of rural women, running since the mid-1990s, has led to a number of new businesses. Leading participants in the Rhön BR have meanwhile joined together to form an alliance for development called "Rhön Quality Products".

Key people, organisational structures, resources

It is evident that in most biosphere reserves, regional marketing has not received sufficient internal or external support. In principle, the higher-level institutions need to be better informed, brought into contact with each other and work together. In most German biosphere reserves to date no special posts have been created to support regional marketing yet. It is mainly the departments dealing with the "greening" of land use that have been given responsibility for this issue.

However, the staff who are responsible for this are often insufficiently qualified (in terms of process and marketing expertise) or do not have the resources they need (working hours, money) to coordinate regional marketing. Cooperation between a state or public institution (biosphere reserve or supporting organisation) and a limited company of stakeholders and their operating businesses has proved to be best practice in some regions.

■ Between consensus and conflict

It has become clear, particularly in the more active regions, that participants who are less innovative (e.g. in agriculture and the food industry, tourism and politics) do not support an over-intensive and sustainability-focused marketing of the biosphere reserves and its products. The influence they can have on the management of the biosphere reserves, to a certain extent in terms of the ecology, quality and image costs, should not be underestimated.

In some biosphere reserves marketing projects are stagnating or failing due to conflicts within the administration. There is a general need for methodological conflict management, supervision and coaching in biosphere reserves.

■ Economy and financing

The business management of regional marketing projects has not yet been investigated scientifically. The profitability of a project, however, always depends on the additional unit costs and returns for the suppliers.

In the case of the successful, independent, smaller projects run by the regional participants, e.g. farmers and restaurateurs in the Rhön BR, the marketing is usually carried out alongside other activities. In these projects the participants seem sufficiently motivated financially to be involved on a permanent basis. To date, no central marketing function has yet been set up in a biosphere reserve.

Outlook

The developments in the Rhön and Schorfheide-Chorin Biosphere Reserves were investigated thoroughly in the above-mentioned R&D Project (Kullmann, A. 2003 c). The results show that the BR administrations should consistently

extend their communication, market and quality leadership, made possible through regional brands. This should also include the surrounding countryside. If they fail to do this, they will be leaving the field open to others who may be competitors or players that do not focus on sustainability.
The trend in regional development is to market a large region internally and externally using the same identity. The image should be the same for both domestic and tourism marketing, in supporting the economy and in regional marketing. The biosphere reserves should try to implement such processes throughout their organisation.
There has hardly been any discussion to date about standard guidelines or characteristics at a national level. "Regional focus plus ecology plus quality" – this combination was described by a regional marketing expert as a formula for the success and survival of his region. This can indeed be applied to most biosphere reserves. With a view to the factors for success, "plus professionalism" should also be added. The stakeholders expect the BR administrations to define concepts clearly and implement them professionally as well as competently. The success factor analysis provides a practical focus for action here.

Literature

KULLMANN, A. (2003 a): Erfolgsfaktoren der Regionalvermarktung. 1. Zwischenbericht zum Forschungs- und Entwicklungsvorhaben "Naturverträgliche Regionalentwicklung durch Produkt- und Gebietsmarketing am Beispiel der Biosphärenreservate". Überarbeitete Fassung. INSTITUT FÜR LÄNDLICHE STRUKTURFORSCHUNG (Ed.), Frankfurt/Main.
KULLMANN, A. (2003 b): Status-Quo der Regionalvermarktung in Biosphärenreservaten. 2. Zwischenbericht zum Forschungs- und Entwicklungsvorhaben "Naturverträgliche Regionalentwicklung durch Produkt- und Gebietsmarketing am Beispiel der Biosphärenreservate". INSTITUT FÜR LÄNDLICHE STRUKTURFORSCHUNG (Ed.), Frankfurt/Main.
KULLMANN, A. (2003 c): Regionalvermarktung in Biosphärenreservaten aus Sicht der wirtschaftlichen Akteure. 3. Zwischenbericht zum Forschungs- und Entwicklungsvorhaben "Naturverträgliche Regionalentwicklung durch Produkt- und Gebietsmarketing am Beispiel der Biosphärenreservate". INSTITUT FÜR LÄNDLICHE STRUKTURFORSCHUNG (Ed.), Frankfurt/Main.
UNESCO (Ed.) (1996): Biosphere Reserves. The Seville Strategy and the Statutory Framework of the World Network, Paris.

5.3 *Integrated Environmental Monitoring – an Ecosystem-Based Approach*

Kati Mattern, Benno Hain and Konstanze Schönthaler

Environmental Monitoring – A Task for Biosphere Reserves

By virtue of their recognition by UNESCO, the German biosphere reserves have entered into an international undertaking to monitor the environment. The "International Guidelines for the World Network of Biosphere Reserves" adopted by the UNESCO General Assembly in 1995 call the biosphere reserves "model regions". There, approaches for the protection of the natural basis of life and sustainable development should be demonstrated at regional level. They should perform this function by supporting environmental monitoring, in particular. The status of environmental monitoring in biosphere reserves is also a component of the review process of the biosphere reserves activities to be conducted every ten years (UNESCO 1996, UNESCO 2002). Within the MAB Programme, Biosphere Reserve Integrated Monitoring (BRIM) sets the framework for environmental monitoring UNESCO biosphere reserves (BRIM 2001).
The German MAB National Committee has included environmental monitoring as a "Functional Criterion" in the "Criteria for the Designation and Evaluation of UNESCO Biosphere Reserves in Germany" (GERMAN NATIONAL MAB COMMITTEE 1996). In the "Guidelines for the Protection, Maintenance and Development" adopted by the *Länder* Working Group on Nature Conservation (LANA) in 1994 the biosphere reserves undertook to contribute to the recording of global environmental problems, such as climate change or the loss of biological diversity (AGBR 1995). This means that there are national requirements and technical voluntary commitments for the BR administrations for environmental monitoring in German biosphere reserves.
The results of environmental monitoring help to meet the obligations from international conventions and decisions, such as Agenda 21. But they are also the basis for checking the success of management schemes taken in the biosphere reserves to implement the protection and development goals. They also provide contributions for reporting to the public (BAYSTMLU/UBA 2000).

Approaches for Monitoring in Biosphere Reserves and Technical Voluntary Commitments for Integrated Environmental Monitoring

Nationwide harmonisation of environmental monitoring in biosphere reserves started in 1993 by developing the 1995 biosphere reserve "guidelines". Until then, the activities of the biosphere reserves in this area were characterised by a "dominance of sectoral monitoring projects". Whereas in some biosphere reserves the conception and establishment of environmental monitoring was still at an early stage, others were already identifying main areas of concentration for monitoring and were setting up permanent monitoring sites (AGBR 1995).

In the 1995 "guidelines" the biosphere reserves agreed on general technical principles for environmental monitoring. Environmental monitoring was to be based on the recommendations of the German Advisory Council on the Environment (*Rat von Sachverständigen für Umweltfragen*, SRU), who identified and defined a need for "General Ecological Environmental Monitoring" in an expert opinion (SRU 1991). In this connection, the monitoring of the environment should be organised at representative sites across all media and sectors, while being adapted to existing time series and sites. It should be more in line with early diagnosis of environmental changes and risk prevention and be based on linking environmental monitoring to ecosystem research and research into effects.

The specific components of the technical voluntary commitments of German biosphere reserves were:

- the harmonised collection of a "core data set" of parameters in all biosphere reserves; this core data set describes structures and processes in the ecosystems and guarantees an integrated, i.e. cross-sectoral environmental monitoring according to the state of the knowledge,
- harmonised task-sharing by space, in which every biosphere reserve is supposed to collect the core data set in selected representative ecosystem types in order to make a contribution to describing the state of the environment in Germany,
- task-sharing by issue, in which the biosphere reserves collect – to secure predictions of trends – additional parameters going beyond the core data set depending on the regional "problems".

The core data set and the proposals for the task sharing of the biosphere reserves were developed in a research project of the Federal Environmental Agency *(Umweltbundesamt/UBA)* (SCHÖNTHALER, K. et al. 1994 and 1997). In the research project "Concept for Integrated Environmental Monitoring – Pilot Project for Biosphere Reserves" a working group of the University of Munich (Prof. Wolfgang Haber) in cooperation with the Berchtesgaden Biosphere Reserve administration made proposals at the conceptual level for the implementation of the requirements of integrated environmental monitoring made in the SRU expert opinion. They gave reasons for the fundamental suitability of biosphere reserves as pilot sites.

The biosphere reserve administrations aimed to implement these technical proposals after they had been more precisely defined – subject to financing by the Federal Government and the *Länder*.

Further Development into a Monitoring Programme Ready for Implementation

In two expert hearings, the scientists involved in environmental monitoring (1995) and the *Land* authorities active in monitoring practice (1997) assessed the programme proposal for integrated environmental monitoring as capable of achieving a consensus (SCHÖNTHALER, K. et al. 1997, UBA 1998). Building on this, the Federal Environmental Agency and the Bavarian State Ministry for Regional Development and Environmental Affairs in cooperation with the environment ministries of Hesse and Thuringia commissioned a further research and development project, which was used to further define and test the concept for integrated environmental monitoring from 1997 to 2001. Bosch & Partner GmbH Munich took lead responsibility together with the Ecology Department of Kiel University, ARSU-GmbH Oldenburg and AG Ökochemie and Umweltanalytik Westerstede for the programme of integrated environmental monitoring. The testing was done in the Rhön BR, which is located in the *Länder* Bavaria, Hesse and Thuringia, under the auspices of the Bavarian BR administration ("Rhön-Project", SCHÖNTHALER, K. et al. 2003). All of the *Land* agencies associated with environmental monitoring in Bavaria, Hesse and Thuringia as well as private facilities that collect data in the Rhön were involved in the project.

The technical information introduced – including information from the biosphere reserves – has played a major role in today's shape of the monitoring programme. The incentives referred to the methodology for deriving the core data set, the interpretability of the monitoring results, the harmonisation of monitoring programmes and the interfaces of integrated environmental monitoring to nature conservation monitoring approaches.

The result is a modular set of integrated environmental monitoring. Its most important elements are:

- **a core data set** of approximately 500 parameters, for which data from existing measurement networks and monitoring programmes for integrated evaluations of environmental problems and of the changes of fundamental ecosystem processes should be made available or collected and/or generated by derivation or modelling. It was developed with the help of a three-pronged approach (problem-based/ system-theoretical-based and data-based). The parameters are assigned to four priority stages. This means that the conditions for a step-by-step implementation have been created. Inquiries on all of the measurement and monitoring programmes conducted in the Rhön BR and its immediate vicinity have shown that the core data set there can be almost completely provided with parameters from the existing programmes and measuring networks.
- **Proposals to create the geographic reference in environmental monitoring**, i.e. procedural proposals to test the degree of representativeness of measuring networks and to generalise monitoring and evaluation results from point to area. The nationwide Site-Ecological Spatial Classification of Germany, drawn up on behalf of the Federal Environmental Agency and the Federal Statistical Office, was used for this (SCHRÖDER, W. et al. 1999 and 2001). In this classification, the Federal Republic is divided into grid cells that are approximately homogeneous in terms of the selected site conditions. The extent to which the grid cells in the

Priority	Parameter groups
Priority 1 257 Parameters	- *Important parameters for the general characterisation of a medium,* - *Parameters for the description of the land use,* - *Most parameters in operating structure and emitter analysis,* - *Parameters for the description of the waterbody structure,* - *The basic chemical and physical parameters (parameters that determine the milieu) for the description of the properties of aqueous solutions and of solid phases,* - *The fractions of the nitrogen, phosphorus and sulphur, where they are to be expected in relevant concentrations in the medium concerned and are important to the substance balance,* - *The (readily available) cations Na^+, K^+, Mg^{2+} and Ca^{2+} in practically all media,* - *The anion Cl^- in practically all media,* - *The heavy metals (in particular the mobile portions) in practically all media,* - *Parameters for identifying the concentrations of oxidisable organic substances and degradable organic substances,* - *Soil microbiological parameters,* - *The most important variables for identifying weather and the climate,* - *Selected biotic parameters,* - *(Passive) reaction and accumulation indicators that react sensitively to the input of eutrophying and acidifying substances and indicate the impacts of heavy metal inputs and* - *Reaction indicators that react sensitively to the influence of photo-oxidants.*
Priority 2 114 Parameters	- *Fractions of the nitrogen, phosphorus and sulphur, where they are expected to occur in low concentrations in the medium concerned and are of less importance to the substance balance,* - *The cations Mn^{2+}, Al^{3+}, Fe^{2+} in practically all media,* - *The total concentration of heavy metals in practically all media,* - *Parameters for the chemical characterisation of waterbody sediments (in particular inorganic substances),* - *Parameters for the chemical characterisation of suspended particles,* - *Organic pollutants in practically all media,* - *Additional variables for identifying weather and the climate,* - *More biotic parameters (e.g. selected species (groups) of soil mesofauna, and* - *Active accumulation indicators that indicate the impacts of heavy metal inputs.*
Priority 3 122 Parameters	- *Fractions of the sulphur, where they are expected to occur in low concentrations in the medium concerned and are of less importance to the substance balance,* - *Organic pollutants in waterbody sediments,* - *Additional variables for identifying weather and the climate,* - *The total concentrations of cations Na^+, K^+, Mg^{2+} and Ca^{2+} in some media (e.g. in the soil solid phase),* - *The cations Mn^{2+}, Al^{3+}, Fe^{2+} in some media (e.g. in precipitation water and/or in runoff from tree trunks and in drops from the treetops),* - *Other biotic parameters (for the vitality and productivity of the plants) and* - *Concentrations of organic pollutants in plant tissues.*
Priority A	- *Parameters for identifying the chemical composition of the air in higher atmospheric layers (N_2O, CH_4, CO_2, CO, O_3)*

Tab.: Priority levels in the core data set, assignment of parameter groups

Rhön Biosphere Reserve are covered in a similar way by monitoring infrastructure was checked. The result was that not all partial areas can be characterised equally well by means of monitoring. This means that single collections of empirical data should be made for checking estimated data generated by models or geostatistical methods.

- **Proposals for the harmonised collection of environmental data** based on the harmonisation efforts of the *Länder* and Federal Government working groups that have been ongoing since the early 1990s, e.g. based on the Guidelines of the *Länder* Working Group on Water "Streams and Rivers in the Federal Republic of Germany – LAWA – Investigation Programme in the *Länder* of the Federal Republic of Germany" (LAWA 1997). Adapting the core data set to the ongoing routine measurement programmes of the Federal Government and the *Länder* key conditions for implementing the integrated environmental monitoring programme have been established.
- **An evaluation concept** as the core element of integrated environmental monitoring, offering opportunities for improving the quality of the findings of the existing monitoring data (evaluation methods that are simpler and already used by the *Land* agencies that are also more complex and that link together different data sets across sectors and media). Associated with this, cause and effect hypotheses on ten environmental problems supposed to be central across Germany were developed. Their structure is based on the indicator systems currently being discussed at international and national level. They define the questions that integrated environmental monitoring is to adopt and form the structure for reporting. The cause and effect hypotheses were regionalised for the Rhön BR in cooperation with representatives from the *Land* agencies and the BR administrations. The regional cause and effect hypotheses with possible future development trends support the emphasis of the data collection on the issues that are particularly relevant to the Rhön BR.
- **A proposal for reporting on integrated environmental monitoring:** To demonstrate the spectrum of possible evaluations, simple as well as complex (model-based) methods for evaluating existing data were used by way of example and the results were prepared for an "exemplary Rhön environmental report". Using the cause and effect hypothesis of "Eutrophication and Acidification of Terrestrial Ecosystems" the report shows the extent to which the use of computer models, as a supplement to the simpler and already tried and tested evaluation methods, is suitable to arrive at more integrated results above and beyond media and sectoral approaches. The environmental report was drawn up solely using the existing data of the *Länder* involved in the Biosphere Reserve. It makes it clear what possibilities for meaningful results exist and can be presented attractively if all institutions that operate monitoring programmes in the Rhön were to bring together their data and design their monitoring programmes jointly so that integrative evaluations of the data would also be possible for all other cause and effect hypotheses.

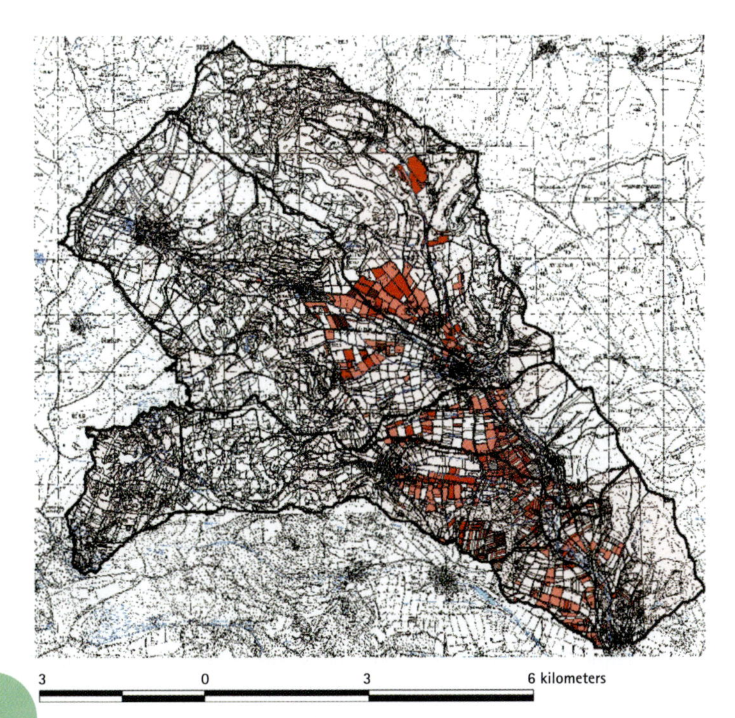

Nitrogen emissions into the surface water seeping through the ground, in 100 cm depth (kg N/ha), actual scenario

border of area
actual N-emissions
surface water (kg N/ha)

0
1-30
31-60
61-90
91-100

Fig.: Model-based evaluation on the risk of nitrogen emissions into the groundwater in the catchment area of the river Streu/Bavarian part of the Rhön Biosphere Reserve (Schönthaler, K. et al. 2003)

Interlinking and Applying Integrated Environmental Monitoring

The result of the research project is that there is a programme for integrated environmental monitoring in the Rhön BR that can be implemented. In 2002, Bavaria, Hesse and Thuringia agreed to continue integrated environmental monitoring and to draw up an environmental report for the Biosphere Reserve on this basis. The manuals and materials (SCHÖNTHALER, K. et al. 2003) provide assistance for comparable activities in other biosphere reserves. Some valuable impetus for the (re)organisation of environmental monitoring programmes was already transmitted during the course of the project thanks to the intensive public relations work of the "Rhön Project".

The quality of the concept of integrated environmental monitoring was elaborately assured by the extensive participation of scientists and *Land* agencies even in the formulation process. The assessments revealed that the monitoring programme is capable of achieving a consensus, both from a scientific and from a technical-administrative point of view. With the help of the "Rhön Project" integrated environmental monitoring, since the publication of the Council's Expert Study, has developed from an initially abstract idea into an implementable instrument (cf SRU 1994, SRU 1998). If an implementation of the system of integrated environmental monitoring is started, the methodological tools according to the state of the art will be available in this concept. If reorganisations in the existing measurement networks of environmental monitoring are due, the concept can help in identifying the parameters that are essential from the systemic point of view.

To enable a gradual implementation of the concept it has a modular structure for implementing each module individually or in succession. This means that the realisation of integrated environmental monitoring can start without additional surveys and without fundamental changes in the administrative structures and organisational processes.

Some biosphere reserves already use single elements of the method modular system, e.g. the Elbe River Landscape BR (Saxony-Anhalt part) and the Vessertal-Thuringian Forest BR. On behalf of the Brandenburg State Institute for Large Protected Areas, Eberswalde University of Applied Science has been developing a concept for integrated environmental monitoring in the Schorfheide-Chorin, Spree Forest and Elbe River Landscape (Brandenburg part) Biosphere Reserves since 1997 parallel to the "Rhön Project" and has already implemented it (LUTHARDT, V. et al. 1999). Repeat surveys are currently being implemented. Close cooperation with the work in the Brandenburg Biosphere Reserves was sought in particular in selecting the parameters for the core data set of integrated environmental monitoring. There was an intensive exchange of results and method developments between the two projects, in particular when drawing up the biotic part of the core data set, including within the context of several workshops. The results show that the approaches concur in numerous aspects and can be complemented. There are also interfaces to the Trilateral Wadden Sea Monitoring (TMAP), which has been ongoing since 1994 and in which the Schleswig-Holstein, Lower Saxony and Hamburg Wadden Sea Biosphere Reserves are incorporated.

When establishing environmental monitoring in the biosphere reserves according to the international MAB requirements, also from the BRIM Programme (HAIN, B. 2001), the biosphere reserve administrations can make use of the method modular system of the "Rhön Project". A brochure with a transparent representation of the monitoring programme in German and English is used to inform the public and those responsible in politics (BAYSTMLU/UBA 2000 and 2002).

The following can be considered for the individual biosphere reserves:

- concluding administrative agreements to secure cooperation with the Federal Government and *Land* agencies involved in environmental monitoring;
- inspecting the cause and effect hypotheses formulated at the national level for their relevance for biosphere reserves and formulating the regionally relevant hypotheses for the area in question;
- bringing together the monitoring activities in the biosphere reserves using the digital questionnaire for documenting metadata developed on behalf of the Federal Environmental Agency (CONDAT GMBH, V. KLITZING, F. 2000). Linking this to the Geographical Information System ("GIS UB") also developed on behalf of the Federal Environmental Agency (SCHRÖDER, W. et al. 2001 and 1999) allows the visualisation and blending of the monitoring programmes with the results of the nationwide site-ecological spatial classification as well as a link to the metadata for describing the programmes. Using a GIS UB questioning module specially developed for the "Rhön Project", those parameters that correspond to the core data set can be selected from the metadata of the monitoring programmes (SCHRÖDER, K. et al. 2001 and 1999).

The following can be considered for all biosphere reserves together:

- checking and updating the desired task-sharing by space and by issue in environmental monitoring, incorporating the Biosphere Reserves that have since been recognised: Upper Lausitz Heath and Pond Landscape (1996), Elbe River Landscape (1997) and Schaalsee (2000). If necessary, the agreed task-sharing by issue must be adjusted to the new circumstances and to changed environmental problems and perceptions of them.

Integrated Environmental Monitoring in the National Context

The contributions for Federal Government and *Land* environmental policy resulting from environmental monitoring should be assessed as just as important as meeting international obligations with respect to the MAB Programme.

The programme of integrated environmental monitoring thus serves the compliance of the decision of the 37th Conference of Environment Ministers of 1991, in which the Federal Ministry for the Environment, Nature and Nuclear Safety (BMU) was asked to further develop environmental monitoring within the meaning of the recommendations of the German Advisory Council on the Environment of 1990.

Environmental monitoring has been legally enshrined at Federal level in the form of Section 12 of the amendment of the Federal Nature Conservation Act (*Bundesnaturschutzgesetz*) (BNATSCHG 2002). Environmental monitoring is thus a task for the Federal Government and the *Länder* within the scope of their responsibilities. They are supposed to support each other in performing this task. GASSNER et al., 2003, have made the following comments on these sections: "specific duties for mutual information, participation in concepts and methods (result) that bring about a division of labour appropriate to the state of affairs and powers". The drawing up of the "Concept for integrated environmental monitoring" on the part of the *Länder* Bavaria, Hesse and Thuringia as well as on the part of the Federal Government within the context of the Rhön BR pilot project for complying with this statutory commission is explicitly cited, "...Changes to the structure and function of the ecosystem concerned shall be recorded as a priority."

At the same time, the modular principle of integrated environmental monitoring means that the programme can also be used for a greater interlinkage of environmental monitoring and environmental reporting, i.e. for forming indicators for policy advice (HAIN, B., SCHÖNTHALER, K. 2003). Methodological and substantial contributions for compliance with existing international, European and national reporting duties (e.g. to the European Environment Agency, EEA) as well as political programmes can be expected from the established integrated environmental monitoring in selected key areas, including biosphere reserves. This is especially the case for reporting duties and/or programmes, which require a greater cross-media and cross-sectoral view of the environment. For example, these include the Federal Government's National Sustainability Strategy (BUNDESREGIERUNG 2001) and the associated activities of the *Länder* to develop core environmental indicators within the context of the Federal Government-*Länder* Working Group on Sustainable Development (BLAK NE) (cf SRU Recommendations 1998, Nos 190-194).

The modular concept of integrated environmental monitoring allows the individual elements also to be used independently for the gradual implementation in "sectoral" monitoring, e.g. for implementation of the reporting duties associated with the EU Flora, Fauna, Habitats Directive or the Convention on Biological Diversity (CBD). The parameters from the Rhön Project, which record the importance of biological diversity for the functioning of ecosystems, are considered to be especially valuable for the Biodiversity Convention. The results of the "Rhön Project" have already been transmitted to the CBD Secretariat and may be included in the activities of the CBD as a "regional case study". In biosphere reserves in particular, the synergetic effects between integrated environmental monitoring and nature conservation related environmental monitoring should be used. The Flora Fauna Habitats Directive makes provision for the recording of substance influences on the species to be monitored. The concept of integrated environmental monitoring makes provision for such collection at the same time and place and for integrative evaluation of biotic and abiotic data as well as of land use data. The use of integrated environmental monitoring for the repeatedly required recording of the state of the environment before the general introduction of genetically modified plants ("baseline") and the associated possible environmental impacts (UBA 2003) is also conceivable.

The involvement of the *Land* agencies means that the technical contents of the "Rhön Project" have already largely been agreed with the *Länder* working groups and the Federal Government-*Länder* working groups. An official information series in the working groups of the Conference of Environment Ministers was started in 2002 with a presentation of the integrated environmental monitoring concept in the Federal Government-*Länder* working group on Environental Information Systems (BLAK UIS) and is continued by the Bavarian Environment Ministry. The aim is to discuss possible strategies for implementation there, especially about the preconditions for an improved data evaluation.

Literature

AGBR (STÄNDIGE ARBEITSGRUPPE DER BIOSPHÄRENRESERVATE IN DEUTSCHLAND) (Ed.) (1995): Leitlinien für Schutz, Pflege und Entwicklung der Biosphärenreservate in Deutschland, Berlin, Heidelberg, New York.

BAYSTMLU, UBA (BAYERISCHES STAATSMINISTERIUM FÜR LANDESENTWICKLUNG, UMWELTBUNDESAMT) (Eds.) (2002): Integrated Environmental Monitoring. Concept and Implementation, München/Berlin.

BAYSTMLU, UBA (BAYERISCHES STAATSMINISTERIUM FÜR LANDESENTWICKLUNG, UMWELTBUNDESAMT) (Eds.) (2000): Ökosystemare Umweltbeobachtung. Vom Konzept zur Umsetzung. Broschüre, München/Berlin.

BNatSchG (Bundesnaturschutzgesetz) (2002): Gesetz über Naturschutz und Landschaftspflege (Bundesnaturschutzgesetz - BNatSchG) vom 25.03.2002 (BGBl I Nr. 22 vom 03.04.2002, p. 1193).

BRIM (Biosphere Reserve Integrated Monitoring) (2001): Report of the Special Meeting on Biosphrere Reserves Integrated Monitoring Office of the Global Terrestrial Observing System (GTOS), Food and Agriculture Organization of the United Nations (FAO) from Rome, Italy, 4 - 6 September 2001.

Bundesregierung (2001): Perspektiven für Deutschland. Entwurf der nationalen Nachhaltigkeitsstrategie, Berlin.

CONDAT GmbH, v. Klitzing, F. (2000): Konkretisierung des Umweltbeobachtungsprogramms im Rahmen eines Stufenkonzeptes der Umweltbeobachtung des Bundes und der *Länder*. Teilvorhaben 1 Überarbeitung des Konzeptes Umweltbeobachtung. Teilvorhaben 2 Fortschreibung der Dokumentation von Programmen anderer Ressorts. Berlin (Umweltforschungsplan des Bundesministeriums für Umwelt, Naturschutz und Reaktorsicherheit. FuE-Vorhaben 299 82 212 / 01, im Auftrag des Umweltbundesamtes).

German MAB National Committee (1996): Criteria for Designation and Evaluation of UNESCO Biosphere Reserves in Germany, Bonn.

Gassner, E. Bendomir-Kahlo, G., Schmidt-Räntsch, A. & J. Schmidt-Räntsch (2003): Bundesnaturschutzgesetz. Kommentar unter Berücksichtigung der Bundesartenschutzverordnung, des Washingtoner Artenschutzübereinkommens, der EG-Artenschutz-Verordnung, der EG-Vogelschutz-Richtlinie und der EG-Richtlinie "Flora, Fauna, Habitat". 2. Auflage, München.

Hain, B. (2001): Concept of Integrated Monitoring and Pilot Implementation in the Rhön Biosphere Reserve. In: UNESCO (Ed.): Special Meeting on Biosphere Reserve Monitoring (BRIM) - Final Report, Paris, pp. 36 - 37.

Hain, B., Schönthaler, K. (2003). Naturwissenschaftliche Anforderungen an Indikatoren. In: Wiggering, H., Müller, F. (Eds.): Umweltziele und Indikatoren. Wissenschaftliche Anforderungen an ihre Festlegung und Fallbeispiele. Gesellschaft für UmweltGeowissenschaften. Berlin, Heidelberg, pp. 141 - 162.

LAWA (Länderarbeitsgemeinschaft Wasser) (Ed.) (1997): Fließgewässer der Bundesrepublik Deutschland, 1. Empfehlungen für die regelmäßige Untersuchung der Beschaffenheit der Fließgewässer in den Ländern der Bundesrepublik Deutschland. 2. LAWA-Untersuchungsprogramm in den Ländern der Bundesrepublik Deutschland, Berlin.

Luthardt V., Vahrson W.-G., Dreger F. (1999): Konzeption und Aufbau der Ökosystemaren Umweltbeobachtung für die Biosphären-reservate Brandenburgs. In: Natur und Landschaft 74 (4), 135-143.

Schönthaler, K., Kerner, H., Köppel, J. & L. Spandau (1994): Konzeption für eine ökosystemare Umweltbeobachtung - Pilotprojekt für Biosphärenreservate. Abschlussbericht zum FuE-Vorhaben 101 04 0404/08 im Auftrag des Umweltbundesamtes, Freising. Unpublished.

Schönthaler, K., Kerner, H., Köppel, J. & L. Spandau (1997): Konzeption für eine ökosystemare Umweltbeobachtung - Wissenschaftlich-fachlicher Ansatz. UBA-Texte-Reihe 32/97, Berlin.

Schönthaler, K., Meyer, U., Pokorny, D., Reichenbach, M., Schuller D. & W. Windhorst (2003): Ökosystemare Umweltbeobachtung; Vom Konzept zur Umsetzung. Bayerisches Staatsministerium für Landesentwicklung und Umweltfragen, Umweltbundesamt (Eds.). In preparation.

Schröder, W., Ahrens, E., Bartels, F. & B. Schmidt (1999): Entwicklung eines Modells zur Zusammenführung vorhandener Daten von Bund und Ländern zu einem Umweltbeobachtungssystem. - 2 Bände, Kiel (FuE-Vorhaben 297 81 126 / 01, im Auftrag des Umweltbundesamtes). Unpublished.

Schröder, W., Eckstein, T., Matejka, H., Pesch, R. & G. Schmidt (2001): Konkretisierung des Umweltbeobachtungsprogramms im Rahmen eines Stufenkonzeptes der Umweltbeobachtung des Bundes und der Länder - Teilvorhaben 3. - Vechta (FuE-Vorhaben 299 82 212 / 02, im Auftrag des Umweltbundesamtes). Unpublished.

SRU (Rat von Sachverständigen für Umweltfragen) (1991): Allgemeine ökologische Umweltbeobachtung. Sondergutachten Oktober 1990, Stuttgart.

SRU (Rat von Sachverständigen für Umweltfragen) (1994): Umweltgutachten 1994. Für eine dauerhaft umweltgerechte Entwicklung. Stuttgart.

SRU (Rat von Sachverständigen für Umweltfragen) (1998): Umweltgutachten 1998. Umweltschutz: Erreichtes Sichern - neue Wege gehen, Stuttgart.

UBA (Umweltbundesamt) (1998): Dokumentation zum 2. Fachgespräch "Konzeption für eine ökosystemare Umweltbeobachtung" vom 20./21. 11. 1997 in Berlin, Fachdokumentation erstellt vom Büro für Ökologie und Planung Göttingen. Unpublished.

UBA (Umweltbundesamt) (2003): Monitoring von gentechnisch veränderten Organismen. Ergebnisse der Modellprojekte von Bund und Ländern, Unterlagen zur Fachtagung vom 27. 5. 2003. Unpublished.

UNESCO (Ed.) (1996): Biosphere Reserves. The Seville Strategy and the Statutory Framework of the World Network, Paris.

UNESCO (Ed.) (2002): Periodic Review for Biosphere Reserves, Paris.

5.4 Socio-Economic Monitoring in the Schleswig-Holstein Wadden Sea Region

Christiane Gätje

Goals of Socio-Economic Monitoring in the Wadden Sea Region ("SEM Wadden Sea")

There are diverse interactions between humans and nature in the Wadden Sea region: people live, work and relax here.

This unique landscape was given national park protection status in 1985 in Schleswig-Holstein, in 1986 in Lower Saxony and in 1990 in Hamburg. Then, in 1990 the Schleswig-Holstein Wadden Sea was recognised as a UNESCO biosphere reserve, followed by the Hamburg and the Lower Saxon Wadden Sea in 1992.

As the Schleswig-Holstein Wadden Sea was first of all a national park and did not become a biosphere reserve until five years later and, furthermore, as the status of a national park entails greater statutory protection, the national park will be used as a unique feature in the elaboration below. But in future, the term "biosphere reserve" will move more to the fore – also due to a planned expansion of the Biosphere Reserve and more recent developments in agriculture (cf Chapter 4.7).

The characteristic natural and cultural landscape of the Wadden Sea Biosphere Reserves forms an important part of the livelihoods for the local population. Three sectors of industry are important here: tourism, fishing and agriculture.

For the majority of large-scale protected areas in Germany (national parks, nature parks and biosphere reserves), monitoring or permanent observation still primarily mean ecological environmental observation. Socio-economic parameters are usually only sporadically part of such programmes.

This is not the case in the Schleswig-Holstein Wadden Sea Biosphere Reserve: between 1989 and 1994 an extensive ecosystem research project initially laid the necessary foundations for the better understanding of structures and dynamics in the Wadden Sea (Stock, M. et al. 1996). The population and economy of the Wadden Sea region, in particular tourism, played a major role in this project from the outset.

The findings were incorporated both in the conception of a trilateral Wadden Sea Monitoring System between Denmark, Germany and the Netherlands, and in the socio-economic monitoring system in the Schleswig-Holstein Wadden Sea Biosphere Reserve, the so-called SEM Wadden Sea.

Benefits of "SEM Wadden Sea"

The "SEM Wadden Sea" has been implemented since 1999 and is used in order to:

- recognise changes and trends in the course of time,
- identify the importance of the protected area for economic development, especially for tourism,
- continuously record the satisfaction of visitors and, where appropriate, increase this satisfaction by means of improved information, visitor guidance as well as protected area and visitor mentoring,
- improve acceptance of the protected area by asking and accommodating the opinions, wishes and interests of the locals and visitors as well as,
- to find new approaches for cooperation and solving conflicts between regional development and development of the protected area.

Contents of "SEM Wadden Sea"

The socio-economic monitoring in the Wadden Sea region comprises three elements (Fig. 1):

SEM Wadden Sea

Elements of Socio-Economic Monitoring

SEM Regional

We are concerned with economic development and the future prospects of the region and want to help shape it.

Statistics and data on
- Population
- Business culture
- Labour market
- Environmental trends

SEM Trend

We want to monitor the trends in visitor numbers, the type and intensity of leisure activities and the expectations and travel motives of holiday-makers.

Conducting
- Censuses
- Visitor surveys
- Estimates
- Mapping

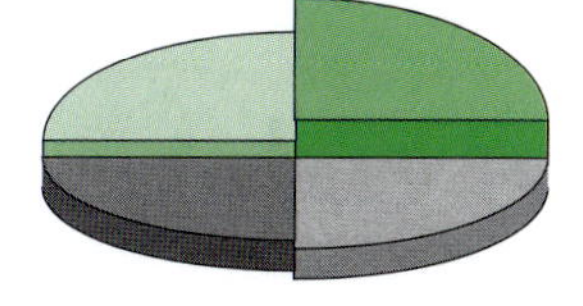

SEM Opinion

We are interested in opinions, wishes, criticism of the people who live here as well as visitors to the North Sea and National Park visitors from all over Germany.

Questioning
- Local inhabitants
- German citizens

Fig. 1: The elements of the socio-economic monitoring in the Schleswig-Holstein Wadden Sea region ("SEM Wadden Sea")

The three modules aim at collecting quantitative and qualitative data as a basis for evaluating social and economic development in the Biosphere Reserve and the area surrounding it.

"SEM Regional" provides the basic data for the economic structural development of the region. Selected data from official statistics are brought together, e.g. for developing tourism in the communities (figures on overnight stays and visitors, length of stay). Supplementary to this, there are individual studies on partial aspects of the economy in the protected area region, e.g. an ongoing project on the regional economic importance of national park tourism.

"SEM Trend" identifies visitor figures and structures as a measurement of the use of the protected area as a site for recreation, leisure activities and environmental education. Furthermore, it also records the level to which the protected area is well known as well as the visitors' satisfaction with the Biosphere Reserve and its offers.

Censuses and short interviews on visitor figures and structures collected at the same time as the visitor survey form another part of the surveys (Tab.).

	1999	2000	2001	2002	Total
Visitors counted[1)]	42,803	27,489	49,492	23,565	143,349
Recording the visitors' structure[2)]	4,089	3,264	5,167	3,200	15,720
Visitor interviews[3)]	572	670	1,019	824	3,085
Survey days	84	116	112	112	424

[1)]Censuses at the 16 survey locations (without interviews)
[2)]Brief interviews, recording visitor type, age, number of adults and children
[3)]Interviews on attitudes, knowledge, expectations and wishes

Tab.: Balance of the visitor surveys in socio-economic monitoring ("SEM Wadden Sea") in the period 1999-2002

When looking at the results of the visitor survey in the Schleswig-Holstein Wadden Sea, the high level of acceptance for protective measures, such as path regulations and entrance restrictions (GÄTJE, C. 2000) as well as for the protective status of the core area as a national park are immediately noticeable.

The visitors to the Wadden Sea are positive, even very positive, about the core area of the Biosphere Reserve, the National Park, as a facility to protect their holiday region: for 68 per cent of the day trippers questioned (n=114) statutory protection of the Wadden Sea was "very important" and for 26 per cent "important". Of 859 holiday-makers at the North Sea, as many as 81 per cent stated that statutory protection of the Wadden Sea was "very important" to them and for another 16 per cent it was "important". The other categories in this survey were "less important/unimportant" and "don't know".

Regular, representative telephone surveys among the regional population to obtain information about the recognition factor, perception and acceptance of the protected area form the third element, "SEM Opinion". These surveys have been conducted since the year 2000 and are supplemented by direct surveys specific to target groups.

The representative survey among locals in 2001, for example, investigated what the regional population associated with the Wadden Sea ("What do you spontaneously think of when you think about the Wadden Sea?").

The term "low tide" (mainly in combination with "high tide") occupied the top position. The species of animal most mentioned was not the Common Seal (*Phoca vitulina*), but rather the Lug Worm (*Arenicola marina*), although this is far from being one of the "sexy species". This creature also appeared often as the second and third choice and seems to embody the Wadden Sea in a symbolic way.

The Lug Worm (Arenicola marina) as a symbolic species for the Wadden Sea. The worm lives in a U-shaped tube in the mud-flats, sandy dropping casts on the sediment surface reveal its presence

Fig. 2 reflects the answers summarised in association fields. It is noticeable that the people questioned most frequently cited characteristic phenomena or elements and/or properties of the Wadden Sea natural area (category "nature that can be experienced"). Furthermore, contemplative terms were also often mentioned (peace, openness, beauty, etc.). Human activities were also among the activities mentioned first – above all walking in the mud-flats. However, they came well behind the diverse natural phenomena (REUSSWIG, F., SCHWARZKOPF, J. 2001).

Overall, economic use played a subordinate role, although this was a survey among the local population.

According to this, the regional population perceives the Wadden Sea as nature that is largely untouched, has been left to its own devices and that visibly regulates itself (the tides), which is mainly accessible. People experience, investigate and enjoy it when walking and perceive it as beautiful (REUSSWIG, F., SCHWARZKOPF, J. 2001), with a feeling of freedom and openness.

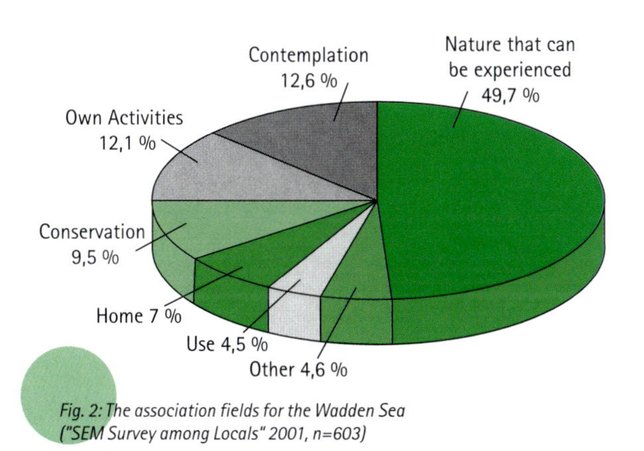

Fig. 2: The association fields for the Wadden Sea ("SEM Survey among Locals" 2001, n=603)

The results of another question, according to which 95 per cent of the locals do not feel personally restricted by the strictly protected areas of the Biosphere Reserve, fit in with this finding.

The locals also appreciate the National Park's biggest information facility, the "Multimar Wattforum", and give it the dream mark of "excellent". 40 per cent of them have already visited the "Multimar Wattforum", half of them several times. Another 29 per cent have planned a visit.

In December 2002 the exhibition was expanded with an extension where, among other things, a complete skeleton of a Sperm Whale (*Physeter macrocephalus*) is displayed. It is to be expected that the opening of this whale house will bring about even more visitors and greater acceptance.

These survey results are a sort of mood barometer for nature conservation in the Schleswig-Holstein Wadden Sea National Park/Biosphere Reserve. They make clear that the protected area meets with great acceptance not only among visitors, but also among the majority of the population.

Transferability of "SEM Wadden Sea"

In German biosphere reserves to date there has been neither an uniform method of collecting data on the figures for visitors' overnight stays and day trippers, nor for the importance of the areas to the regional economy. There are no comparable studies, e.g. on the degree to which the protection status is known or satisfaction and acceptance among locals and visitors.

Establishing a harmonised socio-economic monitoring system as a permanent component of a sustainability monitoring system in UNESCO biosphere reserves could remedy this information deficit. First steps in this direction have already been taken. At a meeting in Rome in 2001 within the context of the project "Biosphere Reserve Integrated Monitoring (BRIM)" of the Programme Man and the Biosphere (MAB), specialists for social monitoring discussed issues, methods, approaches and institutional incorporation relating to monitoring (cf Chapter 3.4). The workshop report (Lass, W., Reusswig, F. 2002) makes specific recommendations, e.g. for guidelines and sets of indicators.

Literature

CWSS (Ed.) (1998): Ministerial Declaration of the Eighth Trilateral Governmental Conference on the Protection of the Wadden Sea. Stade, Germany, October 22, 1997.

De Jong, F., Bakker, J. F., van Berkel, C. J. M., Dankers, N. M. J. A., Dahl, K., Gätje, C., Marencic, H.& P. Potel (1999): Wadden Sea Quality Status Report. Wadden Sea Ecosystem No. 9. Common Wadden Sea Secretariat & Trilateral Monitoring and Assessment Group, Wilhelmshaven, Germany.

Gätje, C. (2000): Der Mensch in der Nationalparkregion. In: Landesamt für den Nationalpark Schleswig-Holsteinisches Wattenmeer (Ed.), Wattenmeermonitoring 1999, Schwerpunktthema: Der Mensch in der Nationalparkregion, Schriftenreihe des Nationalparks Schleswig-Holsteinisches Wattenmeer, Tönning, Sonderheft, pp. 30-51.

Gätje, C. (2003): Socio-Economic Targets for the Wadden Sea. In: Wolff, W. J., Essink, K., Kellermann, A. & M. A. van Leeuwe (Eds.): Challenges to the Wadden Sea. Proceedings of the 10th International Scientific Wadden Sea Symposium, Groningen, The Netherlands, 31 October – 3 November 2000. Ministry of Agriculture, Nature Management and Fisheries/Dept. of Marine Biology, University of Groningen, pp. 221-229.

Gätje, C., Möller, A. & M. Feige (2002): Visitor Management by Visitor Monitoring? Methodological Approach and Empirical Results from the Wadden Sea National Park in Schleswig-Holstein. In: Arnberger, A., Brandenburg, C. & A. Muhar (Eds.) Conference Proceedings "Monitoring and Management of Visitor Flows in Recreational and Protected Areas", January 30-February 02, 2002, Wien, Austria, pp. 68-73.

Hahne, U. (2001): Kommunikations- und Kooperationsstruktur im Nationalpark Schleswig-Holsteinisches Wattenmeer. Study compiled on behalf of the *Land* Agency for the Schleswig-Holstein Wadden Sea National Park. Unpublished.

IRWC (Inter-Regional Wadden Sea Cooperation) (2000): Sustainable Tourism Development and Recreational Use in the Wadden Sea Region. NetForum, Final Report. Ribe County, Denmark.

Landesamt für den Nationalpark Schleswig-Holsteinisches Wattenmeer (2001): Wattenmeermonitoring 2000 – Series of documents by the Schleswig-Holstein Wadden Sea National Park, Tönning, Sonderheft.

Lass, W., Reusswig, F. (Eds.) (2002): Social Monitoring: Meaning and Methods for an Integrated Management in Biosphere Reserves. Report of an International Workshop. Rome, 2-3 September 2001. Biosphere Reserve Integrated Monitoring (BRIM) Series No. 1. UNESCO, Paris. (http://www.unesco.org/mab/brim/workshopdoc/BRIMrept.pdf)

MÖLLER, A. (1996): Socio-Economic Monitoring. In: MARENCIC, H., BAKKER, J., FARKE, H., GÄTJE, C., DE JONG, F., KELLERMANN, A., LAURSEN, K., PEDERSEN, T. F. & J. DE VLAS: The Trilateral Monitoring and Assessment Program (TMAP). Expert Workshops 1995/96. Wadden Sea Ecosystem No. 6. Common Wadden Sea Secretariat & Trilateral Monitoring and Assessment Group, Wilhelmshaven, pp. 66-70.

MÖLLER, A., FEIL, T. (1997): Konzept Sozioökonomisches Monitoring im Nationalpark Schleswig-Holsteinisches Wattenmeer, Report, München/Berlin.

MÖLLER, A., FEIGE, M. (1998): Wirtschaftliche Bedeutung des Tourismus. In: LANDESAMT FÜR DEN NATIONALPARK SCHLESWIG-HOLSTEINISCHES WATTENMEER, UMWELTBUNDESAMT (Eds.): Umweltatlas Wattenmeer, Band 1 Nordfriesisches und Dithmarscher Wattenmeer, Stuttgart, pp. 180-181.

NBV/DWIF (Nordseebäderverband/Deutsches Wirtschaftswissenschaftliches Institut für Fremdenverkehr an der Universität München), (1997): Meer-Wert. Wirtschaftsfaktor Tourismus – Bestandsaufnahme und Perspektiven für die Westküste Schleswig-Holsteins. Husum.

REUSSWIG, F., SCHWARZKOPF, J. (2001): Das Wattenmeer vor Augen – Anmerkungen zum Sozio-ökonomischen Monitoring – Einwohnerbefragung Watt 2001. Potsdam-Institut für Klimafolgenforschung. (Unpublished) Bericht im Auftrag des Landesamtes für den Nationalpark Schleswig-Holsteinisches Wattenmeer.

STOCK, M., SCHREY, E., KELLERMANN, A., GÄTJE, C., ESKILDSEN, K., FEIGE, M., FISCHER,. G., HARTMANN, F., KNOKE, V., MÖLLER, A., RUTH, M., THIESSEN, A. & R. VORBERG (1996): Ökosystemforschung Wattenmeer – Synthesebericht: Grundlagen für einen Nationalparkplan. Schriftenreihe des Nationalparks Schleswig-Holsteinisches Wattenmeer 8.

WOLFF, W. J., ESSINK, K., KELLERMANN, A. & M. A. VAN LEEUWE (Eds.) (2003): Challenges to the Wadden Sea. Proceedings of the 10th International Scientific Wadden Sea Symposium, Groningen, The Netherlands, 31 October – 3 November 2000. Ministry of Agriculture, Nature Management and Fisheries/ Dept. of Marine Biology, University of Groningen, p.p. 7-1. (Recommendations).

5.5 *Allensbach Survey in the Rhön Biosphere Reserve*

Doris Pokorny

Why an Opinion Poll?

The success or failure of the biosphere reserve idea heavily depends on acceptance among the public.

But even ten years after recognition as a UNESCO biosphere reserve in 1991, it was hard to assess what the people in the Rhön really knew about the Rhön Biosphere Reserve and the extent to which they identified with its goals and projects.

Therefore, in the spring of 2002 a professional opinion poll was conducted in the Bavarian, Hessian and Thuringian part of the Rhön Biosphere Reserve by the well-known Allensbach Institute. The questionnaire comprised 43 thematic questions and 13 statistical questions about the individuals. In total, 803 people over the age of 14 were questioned.

Results

The results are very positive and exceeded by far the expectations of the three administrative offices.

72 per cent of those who are familiar with the subject of biosphere reserves associate more advantages with this status, 6 per cent see more disadvantages for the region (Fig. 1). Furthermore, the results demonstrate that the Rhön BR can assume that there is considerable commitment among the population: on a scale of 10 ("the Biosphere Reserve is extremely important to me") to 0 ("the Biosphere Reserve is completely unimportant to me") almost one in five put themselves in the top category. The average classification is 7, in other words surprisingly positive.

However, the Biosphere Reserve is of below average importance to farmers and those involved in agriculture, although

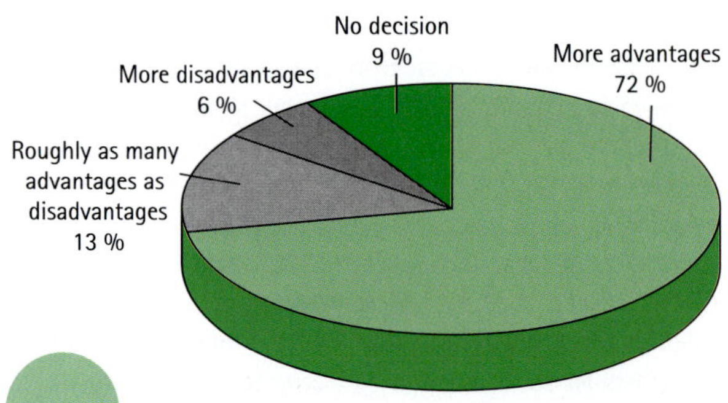

Fig. 1: Biosphere Reserve: Advantages or Disadvantages? (from: HANSEN, J. 2002)

Key ideas about the Rhön: Biosphere Reserve right at the top

Question: "Rhön" can mean many things to different people. I now want to read you a few things that you could think of and you tell me if you think of it when you hear the word "Rhön"

When they hear "Rhön" they think of	Total population	Bavaria %	Hesse %	Thuringia %
a beautiful landscape	99 %	100	99	99
quiet, security	94	96	93	94
Rhön Biosphere Reserve	87	86	86	89
intact, unharmed nature	83	87	83	79
high-quality food from the region	80	72	85	83
pleasant, nice people	76	73	71	88
an economic region with future-oriented	16	10	24	15

Basis: Total population

Fig. 2: Associations with the Rhön (from: HANSEN, J. 2002)

the objective is to preserve the cultural landscape. Also, in response to the question about successes, environmental protection is cited first of all and the marketing of local produce only comes in second place although the Rhön Biosphere Reserve has been very active in this field for many years.

Surprisingly, when they hear the word "Rhön", almost 90 per cent of the inhabitants think of the Biosphere Reserve and one in two said that the name "Rhön Biosphere Reserve" was very familiar to them (47 per cent) (Fig. 2).

The population has a positive view of the information centres, the information material and presentations or guided tours in the Biosphere Reserve. From this, the conclusion can be drawn that the money used for the very expensive public relations work and the diverse print media in the Biosphere Reserve has been invested in the right place.

"Side Effects"

The opinion poll for the first time reached people who so far have not been involved, or not directly involved, as players or representatives of interests in shaping the Biosphere Reserve. The side effects of the opinion poll on local politics must not be underestimated because it has provided proof that there is great acceptance for the Biosphere Reserve among the population.

Although the conditions in the three *Länder* are different (different powers of the administrative authorities, different structural conditions and focuses of work), all questions about the Biosphere Reserve were answered in a similar way by the Rhön inhabitants in Thuringia, Bavaria and Hesse. They all have the impression that the Biosphere Reserve has increased a feeling of belonging among the Rhön inhabitants – a piece of true German unification.

Summary

Above all, the results of the opinion poll provide information about public perception. Where this is not the same as "reality", this is an important indication of shortcomings, e.g. in public relations work and communication.

The same applies in reverse because the administrations of the Biosphere Reserve can have a distorted perception of public opinion.

Opinion polls like these should therefore be conducted at regular intervals as an instrument for cross-checking positions. They would also be suitable as a component of a concept for social monitoring or to monitor success (cf Chapter 3.4).

The results report can be downloaded from the internet in German at www.biosphaerenreservat-rhoen.de (cf "Forschung aktuell").

Literature

INSTITUT FÜR DEMOSKOPIE ALLENSBACH (2002): Biosphärenreservat Rhön – Allensbacher Repräsentativbefragung im Frühjahr 2002. Unpublished.

HANSEN, J. (2002): Das Biosphärenreservat Rhön – aus der Sicht seiner Bewohner, Institut für Demoskopie Allensbach. Unpublished.

5.6 The Schorfheide-Chorin Project: Development of Methods for Integrating Nature Conservation Goals into Agricultural Practice

Eberhard Henne

The Schorfheide-Chorin Biosphere Reserve

In the last few millennia agricultural use of the land by humans has influenced and reshaped most of the natural landscapes on earth. Depending on the conditions of the site, influenced by the climate and limited by the technical scope, primary natural spaces thus developed into a wide spectrum of cultural landscapes (PHILLIPS, A. 1998).

In common practice, today's modern, industrial forms of agricultural production bring about a permanent degradation of natural resources and a dramatic loss of biodiversity (HABER, W. 1986, SRU 1996).

Economy, ecology and social security form an indivisible unit for human economic activity in nature.

Sustainability is achieved only when the long-term security of the natural basis of life are ensured while economic and social living conditions are being improved (BMU 1997).

In this connection, a key role is assigned to cultural landscapes used for agriculture. That is why regional and local standards have to be developed for such areas, containing clear guidelines and specified quality standards on which agricultural use must be based.

From 1994 to 1999 the Federal Ministry for Education and Research (BMBF) and the German Federal Environment Foundation (*Deutsche Bundesstiftung Umwelt* DBU) financed a major multidisciplinary research and development project entitled "Nature Conservation in the Open, Agricultural Landscape using the example of the Biosphere Reserve", for short: the Schorfheide-Chorin Project.

The aim of the project was to develop practical methods and models with which environmental quality goals in the field of nature conservation can be integrated into regular agricultural practise (LEBERECHT, M. 1994).

The characteristics and main focus of the work done by the administration of the Schorfheide-Chorin BR can be summarised as follows:

- bringing together the individual academic results of the work to form a multidisciplinary entirety with an interdisciplinary way of working;
- discussing partial results at different dialogue levels between farmers, academics and administrative staff of the Biosphere Reserve;
- conducting diverse research on a large total area of 1.600 hectares with intensified work on smaller areas and single fields and
- putting great emphasis on the regional character in the research work.

Conducting example projects in the protected area, i.e. harmonised project results were already integrated in practical procedures during the course of the project (FLADE, M. et al. 2003).

The project investigated problems that until then had remained unsolved with existing methods and procedures, in particular:

- influences of land use on the structural elements and the landscape characteristic;
- impacts of individual land use methods on plant and animal communities and the ecosystems;
- methods for how individual partial goals of biotope and species conservation can be formulated as nature conservation quality goals;
- methods for regionalising nature conservation goals;
- drawing up an indicator system for controlling regional nature conservation quality goals;
- influences of nature conservation quality goals on the yields or the crops cultivated in each case;
- results that were achieved with regionalised nature conservation quality goals and, resulting from this, further optimisation of land use methods and
- developing scenario models showing the influence of changing agricultural policy conditions on land use and nature conservation.

As not all nature conservation quality goals could be achieved at the same time in every landscape unit, the goals had to be weighted in the individual basic units of the natural space. Only the priority nature conservation quality goals are considered and incorporated in the proposals for optimisation for land use methods.

In order to ensure that such modifications can be put into practice, so-called field-edge talks were conducted with farmers in the areas of the study.

Three areas were dealt with particularly intensively by all participants in the research. Basic spatial areas in terms of landscape ecology, to which the project results and nature conservation quality goals refer, were separated off.

The aim was to clearly define nature conservation quality goals for the individual agricultural rotations in order to take a decisive step towards future farm models that are kind to nature.

The relevant set of method instruments comprises the following elements:

- Agricultural characteristic and guiding principle

A region's own character results from properties of the location, the spectrum of species and ecosystems as well as from the food and settlement forms. The agricultural characteristic and higher level objectives of nature conservation are the basis of the drawing up of nature conservation guiding principles.

- Quality objective concept

The quality goal methods of technical environmental protection cannot easily be transferred to nature conservation. Due to the natural dynamism and the large number of species, this would lead to a plethora of goals and regulations. According to the current level of knowledge, quality goals can relate only to commodities to be protected and/or certain measures.

- Quality standards and indicators

Since quality goals describe a target state only in terms of quality and are thus not highly suited to implementation in practice, a quantification in the form of quality standards is necessary. Since a direct measurement of quality standards in nature conservation is not possible in many cases, it makes sense to make use of indicators that can be used practically (PLACHTER, H. 1989, RIECKEN, M. 1990). In the individual partial projects, indicators were proposed for the nature conservation quality goals listed and they were then examined in terms of their practical applicability.

- Prioritising quality goals

For all of the quality goals, those were identified that were to have validity in the corresponding basic unit of the natural space (1st prioritisation) and, from these pre-sorted quality goals those are filtered out that are respected on the individual rotation (2nd prioritisation).

- Tolerance threshold model

Fixed thresholds often have a restrictive or counterproductive character for nature conservation quality standards, a way of working aimed at the optimum state would decisively limit the natural dynamism. Optimising models are therefore generally questionable in nature conservation; tolerance threshold models are preferable where thresholds are defined within which certain states or usages are tolerable from a nature conservation point of view.

In the Schorfheide-Chorin Project, a methodological compendium for nature conservation planning in agricultural landscapes and for farms has been developed for the first time:

- transparent regionalisation and definition of nature conservation goals;
- specification of concrete indicators and bundles of measures for nature conservation quality goals;
- consideration of the ecosystem and landscape dynamism;
- models for linking public subsidies to concrete environmental performance and
- support for operational and political decision-making processes.

Although the Schorfheide-Chorin Project had great potential with its overall volume of the four-year research period and the 18-month subsequent treatment, it was obviously not capable of solving all of the problems connected to the application of nature conservation quality goals in cultural landscapes. Nevertheless, the results can make a major contribution to improving sustainable agricultural use if they are taken up by politicians and implemented at the administrative level.

Literature

BMU (BUNDESMINISTERIUM FÜR UMWELT, NATURSCHUTZ UND REAKTORSICHERHEIT) (1997): Auf dem Wege zu einer nachhaltigen Entwicklung in Deutschland. Bericht der Bundesregierung anlässlich der VN-Sondergeneralsversammlung über Umwelt und Entwicklung 1997 in New York, Bonn.

FLADE, M., PLACHTER, H., HENNE, E. & K. ANDERS (2003): Naturschutz in der Agrarlandschaft, Ergebnisse des Schorfheide-Chorin-Projektes, Quelle und Meyer.

HABER, W. (1986): Umweltschutz-Landwirtschaft-Boden. In: Ber. Akad. Naturschutz Landschaftspflege 10, p. 26.

LEBERECHT, M. (1994): Naturschutzmanagement in der offenen agrar genutzten Kulturlandschaft am Beispiel des Biosphärenreservates Schorfheide-Chorin. In: Zeitschrift für Ökologie und Naturschutz 3 (2), pp. 122-125.

PHILLIPS, A. (1998): The Nature of Cultural Landscapes – a Nature Conservation Perspective. In: Landscape Res. 23, pp. 21-38.

PLACHTER, H. (1989): Zur biologischen Schnellansprache und Bewertung von Gebieten. In: Schriftenreihe Landschaftspflege und Naturschutz 29, pp. 107-135.

SRU (RAT VON SACHVERSTÄNDIGEN FÜR UMWELTFRAGEN) (1996): Umweltprobleme der Landwirtschaft (Sondergutachten), Mainz.

RIECKEN, M. 1990: Ziele und mögliche Anwendungen der Bioindikatoren durch Tierarten und Tierartengruppen im Rahmen raum- und umweltrelevanter Planungen. In: Schriftenreihe Landschaftspflege u. Naturschutz 32, pp. 9-26.

WERNER, A., PLACHTER, H. (2000): Integration von Naturschutzzielen in die landwirtschaftliche Landnutzung – Voraussetzung, Methodenentwicklung und Praxisbezug. In: Schriftenreihe Agrarspektrum 31, pp. 44-61.

5.7 Moderation Procedure in the Water Edge Project in the Spree Forest Biosphere Reserve

Elke Baranek, Beate Günther and Christine Kehl

The Spree Forest Water Edge Project is amongst the most interesting projects supported by the "Federal Programme for the establishment and protection of parts of nature and landscapes of national importance" meriting protection in the new *Länder*. The goal of this support is to permanently protect from danger and ultimately improve the ecological quality of sections of extensive, natural and environmentally-friendly landscapes which have prominent national significance.

The project covers an area of approximately 23,000 hectares within the Spree Forest Biosphere Reserve in the *Land* Brandenburg. The planned measures are to be implemented almost exclusively within an area of 8,500 hectares, in the core area and buffer zone of the Biosphere Reserve, which has the status of a nature reserve.

The goals of the project are to maintain and develop the typical features of the Spree Forest with its characteristic range of species, within a cultural landscape that is used sustainably. This includes:

- improvement in the capacity of the landscape to store water,
- revitalisation of low moor land locations,
- improvement in the quality of life in flowing waterbodies and
- development of a land utilisation that is appropriate to the location and spacious areas for succession.

The Water Edge Project had a run-up time of more than ten years. The representatives of public interests and also various users of the land heatedly discussed the initial conceptions in the region as early as 1993.

A joint association was founded from regional administrative bodies and a conservation association in 1998 to act as sponsor for the project. The alliance, represented by the Federal Agency for Nature Conservation (BfN) in Bonn and by the Ministry of Agriculture, Conservation and Environmental Planning of the *Land* Brandenburg, approved the Water Edge Project at the end of the year 2000. For the first time such a large-scale conservation project was drawn up in a procedure with two phases:

- Phase 1: Drawing up of an agreed Maintenance and Development Plan in the region; alongside, specialist planning is supported by a moderation procedure commissioned separately.

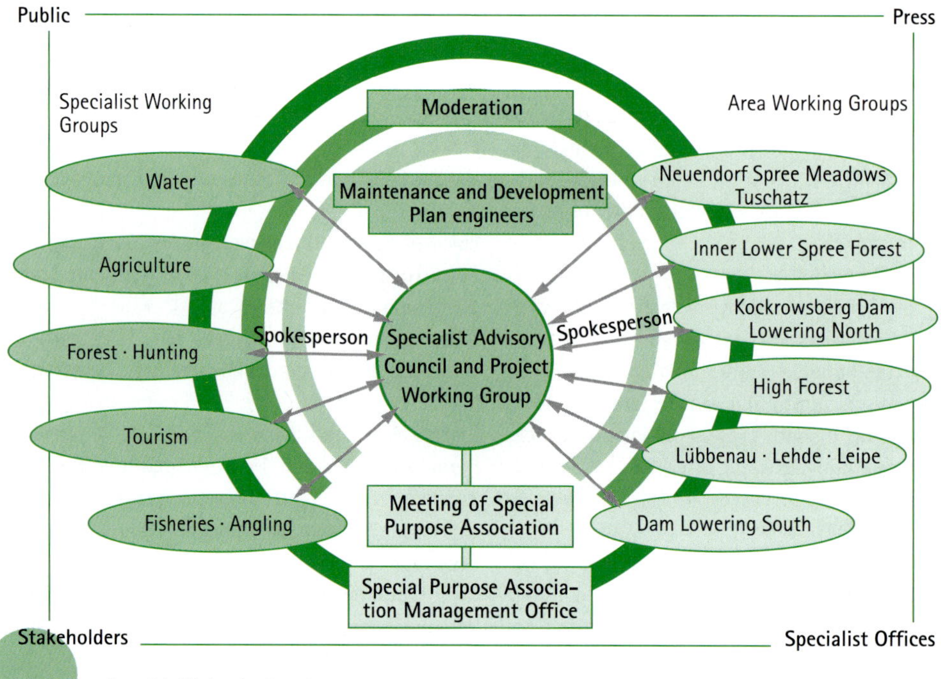

Fig.: An Overview of the Moderation Procedure

- Phase 2: Following the presentation and examination of the Maintenance and Development Plan, the project resources are made available for the implementation of the plans and measures.

In terms of the planning, this process offers greater security to the financial contributors. To safeguard the long-term success of the plans, this corresponds to the philosophy of the UNESCO Man and the Biosphere Programme (MAB) and the Federal Agency for Nature Conservation, which is to work with the citizens and not be in conflict with their interests when implementing conservation measures. The public demands for extensive information and participation will also be dealt with in this way (cf Fig, p. 136).

The moderation procedure began in July 2001 in parallel to the maintenance and development planning and was completed by autumn 2003 with the presentation of the Maintenance and Development Plan.

The goals of the moderation procedure included:

- informing and agreeing on project goals, negotiating opportunities and limits for the large-scale conservation project, explaining its course of action and presenting its results,
- involving regional expertise and experiences in continued ecological investigation as well as in the drawing up of proposals for measures,

Extract from the agreements on cooperation of all participants in the specialist and regional working groups ("The Rules")

1. Agreement about the conditions for action:

- All participants are aware that in the Water Edge Project the funds are to be used primarily for nature conservation in natural and cultural landscapes. When required, alternative possibilities shall be sought for the implementation of further reaching interests and plans.

- Participation shall be effective and different interests shall be included. The results of work and discussions shall be considered within the funding framework. An objective of the specialist and regional working groups shall include technical and regional interests and to discuss them with the engineers, members of the specialist advisory council, the working group supporting the project and project representatives. As a result, recommendations for investigations and measures shall be defined within the framework of the Maintenance and Development Plan for the decision-makers of the joint association and the financial contributors.

- Including the decision-makers early on within the framework of the working group supporting the project etc. shall achieve preliminary agreement that is as comprehensive as possible. However, there is still the possibility that the financial contributors may make a different decision regarding the proposals defined in the Maintenance and Development Plan.

2. Allocation and acceptance of responsibility for all participants, rules for fair behaviour, handling of information:

- Supporters and critics of the project and/or representatives with different interests shall have equal rights as participants in the procedures.

- The common goal shall be to achieve compromises in favour of the Spree Forest region, i.e. to bring about as many win-win-options as possible and, when required, aim for compensatory solutions.

- Discussions shall be taken pragmatically and be result-oriented; opinions must be justified.

- If compromises are not possible, the resulting opinions shall include the views expressed by the minority instead of "KO" votes. These shall be fully justified to the decision-makers.

- When required, decisions relating to procedural matters and the choice of people suitable to act as representatives (speakers, press spokespersons etc.) shall be taken based on the simple majority of those present.

- Everyone shall keep to his or her word, whether in a formal context or over a regular drink after work, i.e. the goal is to achieve internal and external honesty and reliability in discussion and to support compromises.

- Everyone shall have the right to pursue his or her own interests even if they are not in line with the opinions of others.

- The opinions, competencies and experiences of "experts" and "laymen" shall be equally respected; i.e. everyone shall show patience and tolerance and be willing to listen.

- Discussions shall be documented and carried out in such a way as to avoid putting unnecessary pressure on other cooperation initiatives in the region.

- Information and opinions shall be generated by different means: through reporters and spokespersons, through the security of mandates and documentation, etc.

- The type, content and timing of information which is communicated to the working groups internally, in the public eye, or to the management of the joint association, shall be agreed upon. This could be in the form of meetings with the press, press releases, etc.

- As far as is legally and financially possible, information such as concepts for planning, project documents, reports, special conditions and other documentation shall be made available officially and on time.

- The planning for the project and its measures and the decisions to be taken must be comprehensible and intelligible to everyone.

- raising the profile of the interests of various user groups, negotiating, managing conflicts, integrating useful results into the planning and
- discussing and agreeing proposals for measures and related developments in a regional context.

An interdisciplinary moderation team designed and implemented the participation and information process. To establish the course of action to be taken, elements of procedures from environmental mediation were combined with experiences from other information and participation models and project management.

The moderation procedure included the following components:

- a situation analysis: understanding the initial situation, conflict analysis, interviews with representatives of interest groups from the region, summarising the results and drawing up and agreeing rules for cooperation;
- process management: setting up a strategy group; advice and agreement on the overall strategy between representatives of the joint association, the planners, the project management and the moderators;
- public relations work: contacts with the press and media, information events in the region, an exhibition with supporting materials, the development of a project logo;

The burbot (Lota lota) is the central figure of an exhibition and of the Water Edge Ptoject in the Spree Forest

- working groups: setting up regional, area-specific working groups for information exchange and also mediation and technical working groups to advise on general specialist subjects;
- moderation plenum: this was made up of the existing specialist advisory council of the joint association, the working group planned to support the project, the financial contributors and the speakers for the working group. This helped the exchange of information and the preparation of decisions for the joint association meetings;
- the principles for documentation and information: An important factor in the success of the working groups was the transparent preparatory material and careful documentation (minutes) of the results of the discussions.

The content of the discussions mainly focused on:

- the fundamental distrust of all plans, agreements and arrangements that were not handled openly and clearly,
- the stringent demand for complete transparency with regards to action and documentation and
- conflicts between conservation objectives and other uses and different opinions on the relationship between natural and cultural landscapes.

The working groups discussed numerous proposals for measures in several rounds of meetings and finally agreed on the exact wording of the Maintenance and Development Plan. Not all issues and problems could be solved satisfactorily and not all disagreements were cleared up. The information and participation process has, however, made it possible for an extensive plan to be agreed.

Several hundred people from offices, the Biosphere Reserve administration, associations and organisations and land users concerned have worked very intensively on the large conservation project. Without the intensive cooperation in the strategy group, the organisational support of the employees in the project office and last but not least, the commitment of the Spree Forest population, the moderation procedure to support the Maintenance and Development Plan would not have been feasible. The interest of the decision-makers in the region, at a *Land* and Federal level, has emphasised the importance of the Spree Forest Water Edge Project as a very valuable natural and cultural landscape.

The experiences from this moderation procedure are to be analysed and used for other conservation projects. Even if each situation is unique in the beginning and a proposal for information and participation must be devised appropriately for each individual case, many of the process components and the experiences enjoyed here can nevertheless be applied to other regions.

5.8 *Nature Conservation and Organic Farming in a Biosphere Reserve – the Brodowin Eco-Village Development and Testing Project (Schorfheide-Chorin BR)*

Karin Reiter, Johannes Grimm and Helmut Frielinghaus

Organic farming is considered to be a sustainable land use system. Its positive ecological benefits in comparison to conventional and integrated agriculture have been widely documented (STEIN-BACHINGER, K. 1998). However, changing agricultural and economic conditions are heightening the pressure on organic farming, too, to increase yields by means of intensification and to rationalise operational processes. Alongside social aspects and quality demands, environmental protection and nature conservation are still the key motivating forces behind the development and spread of organic farming (HAGEL, I. 2003). Therefore, ways of sensibly integrating aspects of nature conservation into the operational processes of organic farming should be developed.
The Brodowin Eco-Village is a large "Demeter" farm (1,239 hectares) in the midst of the Schorfheide-Chorin Biosphere Reserve (cf box). The poor quality of the soils and the associated relatively low cropping intensity offer good conditions for the integration of nature conservation objectives (STEIN-BACHINGER, K. et al. 2002, FUCHS, S. et al. 2003, STEIN-BACHINGER, K. et al. 2003). In close cooperation with the farm, a development and testing project of the Federal Agency for Nature Conservation (BfN) is evolving and testing cultivation methods (in practice under real working and market conditions) that have been optimised in terms of nature conservation. The measures are also evaluated economically at farm level. The results are incorporated in a farm organisation that has been expanded by nature conservation goals. For example, in cereal farming the intensity of weed control is reduced and the proportion of spring crops is increased. In legume-grass forage the increase in cutting height, delayed cutting periods and the use of special techniques are being tested. The arrangement of crop rotation fields is changed according to aspects of nature conservation; furthermore, the field structures are optimised by limiting the field size to 25 hectares and creating additional structures.
The impacts of the changed methods are examined on the basis of selected indicator species of the following groups: farmland and hedgerow birds, amphibians (*Amphibia*), butterflies and moths (*Lepidoptera*), grasshoppers (*Saltatoria*) as well as the indicator species Brown Hare (*Lepus europaeus*) and segetal flora/arid grassland vegetation (cf www.naturschutzhof.de).

The Skylark (Alauda arvensis) is one of the species of fauna under observation in Brodowin.

Brodowin Eco-Village – Organic Farms in a Biosphere Reserve

The "Brodowin Eco-Village" farm has been using bio-dynamic methods since 1990. Surrounded by seven lakes, Brodowin lies in the Schorfheide-Chorin Biosphere Reserve in the Uckermark (cf www.brodowin.de). The farmland is 60 m above sea level; the average annual precipitation is 500 mm, the average valuation index of the fields is 33 (the valuation of the field is an evaluation measure for the soil-quality, from 7 (poor) up to 100 points (high quality)).
At 1,239 hectares Brodowin is one of the biggest organic farms in Germany. 1,167 hectares are arable land, 47 permanent grassland, 25 are available for growing vegetables and 8 for permanent crops. In the crop rotation, a mixture of alfalfa (Medicago sativa subsp. sativa) and grass or clover (Trifolium pratense) and grass, Common Wheat (Triticum aestivum), Spelt Wheat (Triticum spelta), peas (Pisum sativum subsp. sativum), rye (Secale cereale) and other cereals are grown. Thirty varieties of vegetables and herbs are cultivated in the open and in 2,500 square metres of greenhouses.
The dairy cow herd is the heart of the farm. 290 dairy cows are kept with 350 young stock (Holstein-Friesian crossbreeds). The cattle have the best possible living conditions on the pasture and, in the winter, on straw bedding in large loose housing. The raw milk – 4,500 litres every day – is processed in the farm's own dairy. The product range includes market milk, butter, artisan cheese and mozzarella. The Brodowin Eco-Village "Demeter" Farm sees itself as a regional supplier for the Greater Berlin area. The farm shop and delivery service reach 1,600 families. The farm is a demonstration farm within the Federal Organic Farming programme.
(http://demonstrationsbetriebe.oekolandbau.de)

The data collected to date suggest that changes to crop rotation (increasing the proportion of spring cereals, legume-grass forage) clearly increase the number of species and the population density of typical farmland birds (FUCHS, S. et al. 2003). As well as optimising arable farming methods, crop rotation and field structures, the establishment of an ecological network of grazed arid grassland is also planned. Moreover, land management waste (e.g. hedge clippings) are composted and used to improve the soil structure and the carbon balance of the farm.

Passing on the project idea and conveying the results to the public play an important role. The project has been presented at many events, such as farm festivals, guided tours, agricultural events (e.g. International Green Week in Berlin) or scientific conferences.

Observing the Skylark (Alanda arvensis) in Brodowin

Indicator Species of the Development and Testing Project:

Farmland birds:	*Skylark (Alauda arvensis), Corn Bunting (Miliaria calandra), Yellow Wagtail (Motacilla flava), Whinchat (Saxicola rubetra), Quail (Coturnix coturnix) and Grey Partridge (Perdix perdix)*
Hedgerow birds:	*Red-Backed Shrike (Lanius collurio) and Barred Warbler (Sylvia nisoria)*
Mammals:	*Brown Hare (Lepus europaeus)*
Amphibians:	*European Fire-Bellied Toad (Bombina bombina), Tree Frog (Hyla arborea) and Common Spadefoot (Pelobates fuscus)*
Insects:	*butterflies and moths (Lepidoptera) and grasshoppers (Saltatoria)*
Plants:	*segetal flora and arid grassland vegetation*

Literature

FUCHS, S., GOTTWALD, F., HELMECKE, A. & K. STEIN-BACHINGER (2003): Erprobungs- und Entwicklungsvorhaben "Naturschutzfachliche Optimierung des großflächigen Ökolandbaus am Beispiel des Demeter-Betriebes Ökodorf Brodowin". In: BUNDESAMT FÜR NATURSCHUTZ (Ed.): Treffpunkt Biologische Vielfalt III, Bonn, pp. 97-102.

HAGEL, I. (2003): Zu einer Weiterentwicklung des Qualitätsbegriffes im ökologischen Landbau. In: INSTITUT FÜR BIOLOGISCH-DYNAMISCHE FORSCHUNG E. V. (Ed.): Annual Report 2002, pp. 41-45.

STEIN-BACHINGER, K. (1998): Leistungen und Potenziale des ökologischen Landbaus für den biotischen und abiotischen Ressourcenschutz. In: Conference Reports from the 4th Nature Conservation Conference of the Brandenburg Land Association of NABU: Arten- und Ressourcenschutz in der Landwirtschaft, pp. 27-40. NABU-Brandenburg@t-online.de (Ed.).

STEIN-BACHINGER, K., BACHINGER, J., FUCHS, S. & P. ZANDER (2002): Managementsysteme von Ackerflächen des ökologischen Landbaus zur Integration naturschutzfachlicher Ziele. In: Mitt. Ges. Pflanzenbauwiss. 14, pp. 121-122.

STEIN-BACHINGER, K., ZANDER, P. & S. FUCHS (2003): Optimierung des ökologischen Landbaus auf Grundlage naturschutzfachlicher und betriebswirtschaftlicher Aspekte. In: FREYER, B. (Ed.): Beiträge zur 7. Wissenschaftstagung zum ökologischen Landbau, Verlag Universität für Bodenkultur, Institut für ökologischen Landbau, Wien, pp. 165-168.

STEIN-BACHINGER, K., ZANDER, P. & S. FUCHS (2003): Optimisation of Organic Agriculture on the Basis of Nature Protection and Economic Aspects. English Abstracts (CD) of the 7th Scientific Conference of Organic Agriculture in Vienna: The Future of Organic Agriculture. Verlag Univ. für Bodenkultur, Institut für ökologischen Landbau, Wien.

5.9 Further Development of the "Ecosystem Approach" of the Convention on Biological Diversity in Selected Forest Biosphere Reserves

Anke Höltermann

By drawing up and implementing appropriate measures (Art. 6 CBD), the countries that have signed the Convention on Biological Diversity (CBD) commit themselves to the conservation and sustainable use of biological diversity for the benefit of people today and future generations. The fundamental objective of the CBD is the protection of biodiversity, the sustainable use of biological resources and benefit sharing when using genetic resources. The Convention was ratified by Germany in 1993 and came into force on 21 March 1994.
The so-called ecosystem approach is an important innovative initiative within the CBD. The concept of the CBD ecosystem approach was put into concrete terms for the first time in 1998 with the so-called Malawi Principles. In the year 2000, with the slight changes made at the fifth Meeting of the Parties to the CBD held in Nairobi, it was recommended that the signatory countries begin their implementation. The boxes on the right and on the next page provide an overview of the twelve principles and five guidelines of the CBD's Decision V/6.
Against a backdrop of fundamental uncertainty that exists when dealing with complex, non-linear systems such as ecosystems, the ecosystem approach, among other things, calls for the development of integrative management strategies using adaptive management methods (see Häusler, A., Scherer-Lorenzen, M. 2002: 11). These are to coordinate the three objectives of the CBD: use and protection of biodiversity and benefit sharing when using genetic resources. The management objectives should be agreed by all the relevant social groups and should maintain an appropriate balance between the objectives of the CBD.
With the ecosystem concept of the CBD, the original, more abstract and scientifically oriented ecosystem concept has been expanded to include social, administrative, political and economic dimensions of resource management. As an action-oriented political component, it has become the shorthand for a "holistic" approach, which aims for the management of ecosystems to span the media and institutions.

Principles of the Ecosystem Approach

Principle 1: *The objectives of management of land, water and living resources are a matter of societal choice.*
Principle 2: *Management should be decentralized to the lowest appropriate level.*
Principle 3: *Ecosystem managers should consider the effects (actual or potential) of their activities on adjacent and other ecosystems.*
Principle 4: *Recognizing potential gains from management, there is usually a need to understand and manage the ecosystem in an economic context. Any such ecosystem-management programme should:*
a) reduce those market distortions that adversely affect biological diversity;
b) align incentives to promote biodiversity conservation and sustainable use;
c) internalize costs and benefits in the given ecosystem to the extent feasible.
Principle 5: *Conservation of ecosystem structure and functioning, in order to maintain ecosystem services, should be a priority target of the ecosystem approach.*
Principle 6: *Ecosystems must be managed within the limits of their functioning.*
Principle 7: *The ecosystem approach should be undertaken at the appropriate spatial and temporal scales.*
Principle 8: *Recognizing the varying temporal scales and lag-effects that characterize ecosystem processes, objectives for ecosystem management should be set for the long term.*
Principle 9: *Management must recognize that change is inevitable.*
Principle 10: *The ecosystem approach should seek the appropriate balance between, and integration of, conservation and use of biological diversity.*
Principle 11: *The ecosystem approach should consider all forms of relevant information, including scientific and indigenous and local knowledge, innovations and practices.*
Principle 12: *The ecosystem approach should involve all relevant sectors of society and scientific disciplines.*

Operational guidance for application of the ecosystem approach

1. *Focus on the functional relationships and processes within ecosystems*
2. *Enhance benefit-sharing*
3. *Use adaptive management practices*
4. *Carry out management actions at the scale appropriate for the issue being addressed, with decentralization to lowest level, as appropriate*
5. *Ensure intersectoral cooperation*

An essentially comparable focus is pursued by the concept of the UNESCO Man and the Biosphere Programme (MAB). The objective of the Seville Strategy is to use biosphere reserves as models for sustainable development (UNESCO 1996). With its worldwide network of representative natural and cultural landscapes, the UNESCO biosphere reserves have over 440 areas worldwide that are suitable for investigation. This provides an opportunity for cooperation.

On the occasion of the sixth Meeting of the Parties to the CBD in The Hague (2002), the ecosystem approach was also incorporated into the CBD's "Extended Working Programme for Forests" (cf Decision VI/22).

In line with Programme Element 1, the first objective of the programme should be to apply the ecosystem approach to the management of all types of forests. The contracting parties, including Germany therefore, are required to draw up regional guidelines for the application of the ecosystem approach and to review these within the framework of case studies. What is more, by 2006 there should be a revision of the twelve criteria on the basis of the case studies and experiences gathered by then.

However, there are various difficulties in this respect. For example, when expressing the concept in concrete terms for individual cases, there is great scope for interpretation due to the rather abstract wording of the twelve principles and guidelines. This prevents the concept being applied directly in practice (cf Korn, H. et al. 2003). On the one hand, the flexibility in the interpretation of the concept is the reason for its popularity. On the other hand, however, it is restrictive when applied to individual cases.

Furthermore, in the specific case of forests, the lack of clarity in applying the concept for sustainable use to forest areas makes it difficult to define the specific requirements for implementing the ecosystem approach. Recognising this fact, CBD's Decision VI/12 2(b) calls for clarification of the differences and overlaps of the ecosystem approach with the approach for "sustainable forest management". The first investigations on this are provided by Ellenberg, H. (2003) and Häusler, A. et al. (2002).

Biosphere reserves can play an important role both in defining the initial stages of further development for the ecosystem approach and in implementing the programme elements of the "Extended Working Programme for Forests in Germany". Research approaches resulting from questions raised and consolidated within the framework of a research project supported by the Federal Agency for Nature Conservation (BfN) since 15.08.2003, ("The 'Ecosystem Approach' in Selected Biosphere Reserves") will be discussed in more detail below.

UNESCO biosphere reserves are model regions, where the priorities are the preservation of natural and cultural landscapes, the strengthening of regional economies, the inclusion of the population in establishing living, working and recreation areas and research and education. They aim for economic and cultural development and improvement of cultural landscapes without destroying the natural foundations of life. They can – in the special case of biosphere reserves consisting mainly of forests – make an important contribution to the further development of adaptive and integrative conservation strategies in forest areas.

Within the framework of the current BfN research and development project mentioned above, the *Forschungsanstalt für Waldökologie und Forstwirtschaft* (Research Institute for Forest Ecology and Forestry) in Trippstadt, in cooperation with the universities of Freiberg and Kaiserslautern, will use empirical studies to review how the principles and guidelines of the CBD ecosystem approach can be applied to biosphere reserves. Real problems and deficiencies with the implementation of the criteria should be identified in the process and possible proposals for solutions developed. With the help of the relevant interest groups in the area, strategies and methods should be developed to express the principles more precisely. By including various biosphere reserves found in representative cultural or natural landscapes in Germany and in different general socio-economic conditions, the significance and applicability of the results can be taken beyond the existing model region.

Based on the results, the recipients of the research plan to develop ideas and proposals for the discussion and possible revision of the 12 principles of the ecosystem approach. These can then be included in the international discussion process in the run up to the eighth Meeting of the Parties 2006. Despite the exclusively national focus and the different ways in which the ecosystem approach has been implemented in each area of investigation, the ability to put this into practice at an international level and the corresponding wording of the revision proposals is a particular priority in this phase.

As a final step, the Research Institute for Forest Ecology and Forestry intends to draw up final proposals for the formal and institutional establishment of an international network of model regions, with exemplary implementation of the ecosystem approach. In this context, biosphere reserves could also lead by example in going beyond the mere conception of further development and demonstration of the ecosystem approach. If during the course of the research biosphere reserves prove to be suitable model regions for the demonstration of the ecosystem approach, there will also need to be an investigation at an international level of the integration of biosphere reserves into a network of model regions.

Literature

ELLENBERG, H. (2003): "Ecosystem Approach" versus "Sustainable Forst Management" – Versuch eines Vergleichs. Arbeitsbericht der Bundesforschungsanstalt für Forst- und Holzwirtschaft, Universität Hamburg.

HÄUSLER, A., SCHERER-LORENZEN, M. (2002): Nachhaltige Forstwirtschaft in Deutschland im Spiegel des ganzheitlichen Ansatzes der Biodiversitätskonvention, BfN-Skripten 62, Bonn.

KORN, H., SCHLIEP, R. & J. STADLER (2003) (Eds.): Report of the International Workshop on the "Further Development of the Ecosystem Approach", BfN-Skripten 78, Bonn.

UNESCO (Ed.) (1996): Biosphere Reserves. The Seville Strategy and the Statutory Framework of the World Network, Paris.

Research project "Ecosystem Approach in Selected Forest Biosphere Reserves"

Contact:

Dr. Ulrich Matthes, Forschungsanstalt für Waldökologie und Forstwirtschaft Rheinland-Pfalz; Tel: +49 6306/911-153; e-mail: Ulrich.Matthes@wald-rlp.de

Dr. Anke Höltermann, BfN; Tel: +49 228/8491-417; e-mail: HoeltermannA@bfn.de

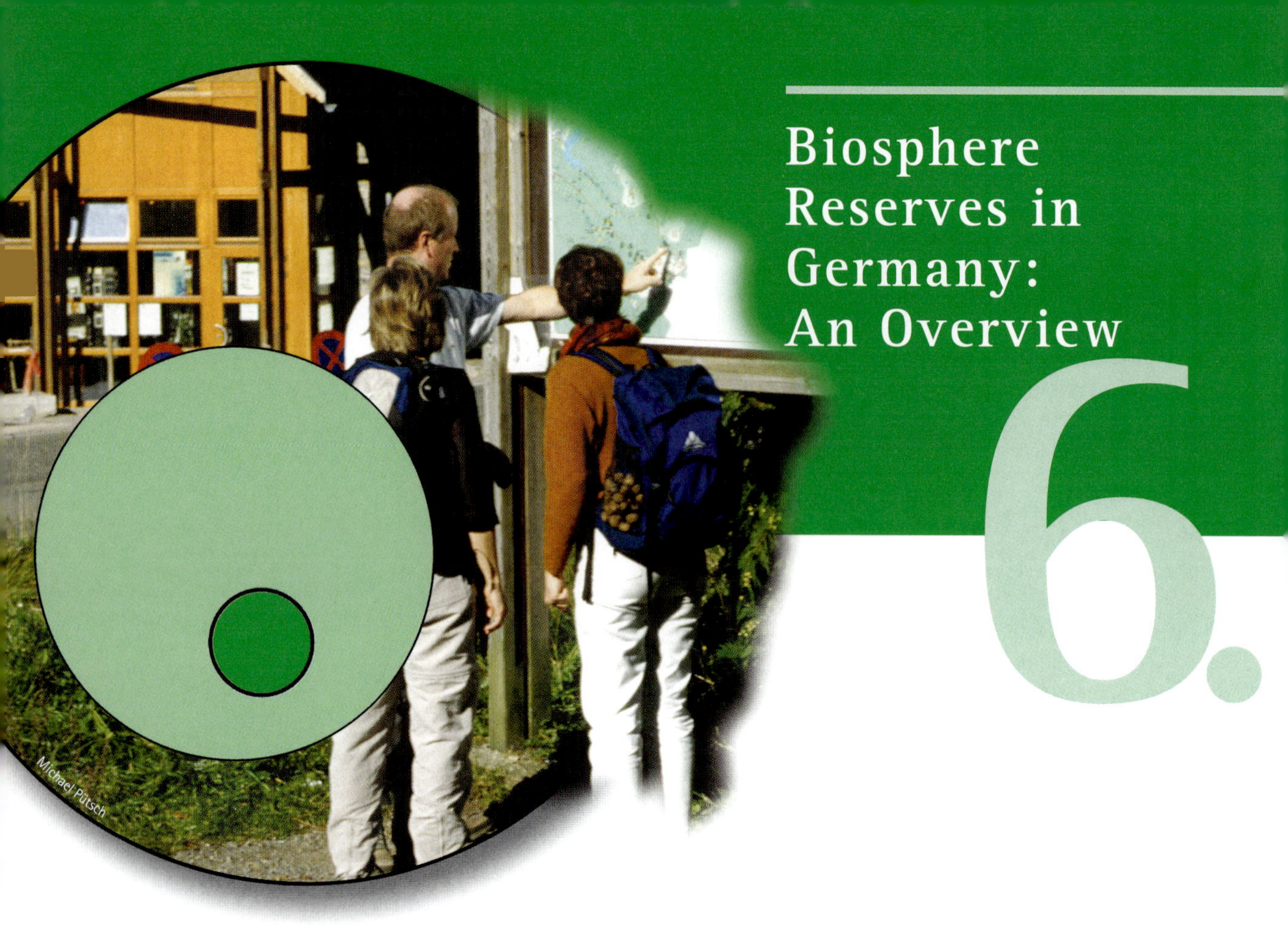

Biosphere Reserves in Germany: An Overview

6.

Name
Federal Republic of Germany (FRG)

International codes
D, DE, GER

Location
Central Europe

Capital city
Berlin with a population of 3,386,667 (2000)

Official language
German

Area
357,022 square kilometres (world ranking 61)

Population
Inhabitants:
82,150,000 (average 230 per km²); most highly populated country in the EU (world ranking 12); 91.1 % Germans, 8.9 % foreigners; minorities with special rights: Sorbs (Wends), Danes, Friesians, Sinti and Roma

Refugees (late 2001):
116,000 from Yugoslavia and Bosnia-Herzegovina

Life expectancy (2000): *77 years*

Infant and child mortality (2000): *0.5 per cent*

Population growth (average 1980–2000): *0.2 per cent*

Illiteracy rate (1998): *below 5 %*

Religion:
Roman Catholic Church (1999): 27,017,000; Protestant Church (1999): 26,800,000; Islam (2001): 3,200,000; and others

State
The Federal Republic of Germany is a federal state and consists of 16 Federal Länder: Baden-Württemberg [BW], Bavaria [BY], Berlin [BE], Brandenburg [BB], Bremen [HB], Hamburg [HH], Hesse [HE], Mecklenburg-Vorpommern [MV], Lower Saxony [NI], North Rhine-Westphalia [NW], Rhineland Palatinate [RP], Saarland [SL], Saxony [SN], Saxony-Anhalt [ST], Schleswig-Holstein [SH] and Thuringia [TH]

The eleven Länder of the original Federal territory (BW, BY, BE, HB, HH, HE, NI, NW, RP, SL, SH) were refounded or created after 1945. After the first free elections in the German Democratic Republic (GDR) on 18 March 1990 the parliamentarians in the Volkskammer decided to create five Federal Länder (BB, MV, SN, ST, TH). On 3 October 1990 the accession of the GDR to the purview of the Basic Law of the FRG was completed; since then 3 October has been the German National Day: The Day of German Unity.
Parliamentary democratic Federal state since 1949; Basic Law from 1949; elections every four years; the Bundestag represents the people and the Bundesrat represents the Länder; suffrage over the age of 18.
Each Federal Land has its own Land Constitution, Land Parliament and Land Government. Federal responsibility in exclusive (e.g. foreign affairs), competing (e.g. civil and criminal law) and general legislation (e.g. nature conservation and landscape management).

Parties:
Sozialdemokratische Partei Deutschlands (SPD) [Social Democrats], Christlich-Demokratische Union – Christlich Soziale Union (CDU/CSU) [Christian Democrats and Christian Social Union], Bündnis 90 / Die Grünen (B'90/Grüne) [Alliance 90/The Greens], Freie Demokratische Partei (FDP) [Free Democrats], Partei des Demokratischen Sozialismus (PDS) [Party of Democratic Socialism], and others

Economy
GNP (2000) $2,063,734 million; GDP (2000): $1,872,992 million; breakdown: agriculture 1.2 %, industry 30.1 %; services 68.7 %

Further Information
www.bundesregierung.de/en

Source: Federal Agency for Nature Conservation (BfN), 2003

Biosphere reserves

Settlement areas

6.2 UNESCO Biosphere Reserves in Germany

The following pages give an overview over important facts and figures on the 14 biosphere reserves in Germany.

Impressions from the German biosphere reserves

South-East Rügen Biosphere Reserve

Schleswig-Holstein Wadden Sea Biosphere Reserve

Hamburg Wadden Sea Biosphere Reserve

Lower Saxon Wadden Sea Biosphere Reserve

Schaalsee Biosphere Reserve

Schorfheide-Chorin Biosphere Reserve

Elbe River Landscape Biosphere Reserve

Spree Forest Biosphere Reserve

Upper Lausitz Heath and Pond Landscape Biosphere Reserve

Vessertal-Thuringian Forest Biosphere Reserve

Rhön Biosphere Reserve

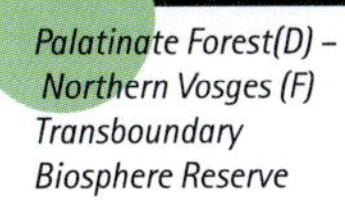

Palatinate Forest(D) – Northern Vosges (F) Transboundary Biosphere Reserve

Bavarian Forest Biosphere Reserve

Berchtesgaden Biosphere Reserve

Culture, Landscape, Sea and More

South-East Rügen Biosphere Reserve

Federal Land
Mecklenburg-Vorpommern (MV)

Year of UNESCO Recognition
1991

Administration
Rügen National Park Office
Blieschow 7a
D-18586 Lancken-Granitz
Tel.: +49 38303/885-0
Fax: +49 38303/885-88
e-mail: info@nationalparkamt-ruegen.de

Director: *Dr Michael Weigelt*

Number of full-time employees: *24.5 (out of a total of 54)*

Competent Authority
Mecklenburg-Vorpommern Ministry for Food, Agriculture, Forestry and Fisheries (supreme authority), Mecklenburg-Vorpommern Environment Ministry (nature conservation specialist supervision), Schwerin

Information Centre
None

Information Material
"... das wahre Paradies von Rügen" ["... the true paradise of Rügen"] (basic flyer), "Baustilfibel Rügen" ["Rügen architectural guide"], various activity reports

Homepage
www.biosphaerenreservat-suedostruegen.de

Area
Total:
23,500 ha (= 12,600 ha Bodden (lagoons) and Baltic Sea, 10,900 ha land)
Ownership: cf information on zones

Core area:
349 ha (200 ha land, 149 ha water)
Ownership: 34.7 % Federal Government, 25.8 % Land, 25.5 % "East Rügen" landscape management association, 2.3 % BVVG, 11.7 % private

Buffer zone:
3,204 ha (1,354 ha land, 1,850 ha water)

Transition area:
19,947 ha (8,993 ha land, 10,954 ha water)
Ownership (buffer zone and transition area): 1.5 % Federal Republic, 23.4 % BVVG, 0.5 % TLG , 9.4 % Land, 0.1 % Mecklenburg-Vorpommern Land company, 1.5 % Putbus Council, 5.9 % municipalities, 5.8 % "East Rügen" landscape management association, 0.1 % Deutsche Bahn AG, 51.9 % private

Geographical Location
south-easter Rügen with partial areas of Granitz, Mönchgut, area surrounding Putbus, Vilm Island, northern part of the Rügen Bodden

Nature Space and Ecosystem
representative section of the Bodden coast with late ice age island cores, steep banks and sandy beaches

Population in the Biosphere Reserve
11,600 (106 per km², only related to land area)

Fulfilment of the Functions
Conservation:
responsibility as lower nature conservation authority; contractual nature conservation (grassland extensification); various renaturalisation projects (especially coastal flooding moors); measures for species and biotope protection; landscape management; visitor guidance; agreements with water users; management and development plans; concepts, specialist reports, etc. Federal sponsoring projects for areas with nationally representative significance "Eastern Rügen Bodden Landscape"

Development:
moderated implementation concept "Rügen Model Region" (1998/99, 5 sub-projects, 17 working groups); "Biosphere Job Motor" (since 1999, www.job-motor.de); "Junior Biosphere Job Motor" (since 2002); "Biosphere Marketplace" (since 2002); "Posewald Project School" (since 1996); "Biosphere Ticket" Rügen (since 2002); various umbrella and individual concepts for the regionalisation of agriculture and fisheries on Rügen; participation in projects in the district (LEADER)

Logistic Support:
public relations work and environmental education (many different activities); cooperation with schools; "Rügen Wood and Regional Fair" (since 1997, www.ruegener-holzmesse.de); "Biosfestival" ("blue boat" International Youth Jazz Festival and cooperation with the "Putbus Festival"); cooperation agreement with Mittweida University of Applied Science and seven more universities of applied science since 1993, cooperation with various universities; traffic monitoring (since 1993); species monitoring

Regular Events
"Rügen Wood and Regional Fair"; "blue boat" International Youth Jazz Festival; start-up days and "regular meetings" in the "Biosphere Job Motor"

Sponsoring Agency
"Förderverein Modellregion Rügen e. V." Rügenhaus,
Binzer Str. 50
D-18528 Zirkow
Tel.: +49 38393/133829
e-mail: foerderverein@modellregion-ruegen.de

Partnerships
Wollin National Park (Poland), Columbian Central National Park Administration, Archipelago Sea BR (Finland), Vilsandi National Park (Estonia), contacts with Denmark and Australia

Special Features
Rügen National Park Office competent Land authority for the South-East Rügen BR (lower nature conservation authority) and for the Jasmund National Park (lower forest and nature conservation authority)

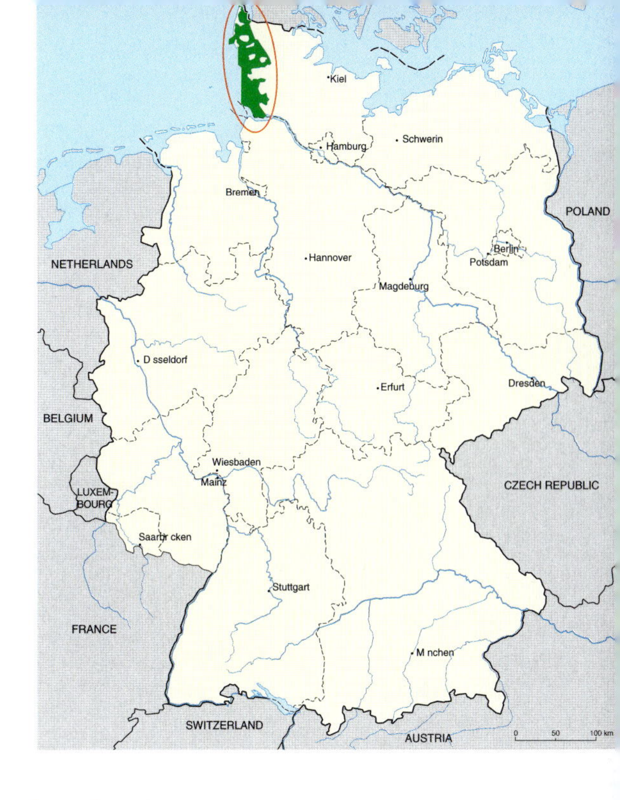

Where the Sea Floor Meets the Horizon

Schleswig-Holstein Wadden Sea Biosphere Reserve

Federal Land
Schleswig-Holstein (SH)

Year of UNESCO Recognition
1990

Administration
Landesamt für den Nationalpark
Schleswig-Holsteinisches Wattenmeer
Schloßgarten 1
D-25832 Tönning
Tel.: +49 4861/616-0
Fax: +49 4861/616-69
e-mail: info@nationalparkamt.de

Director:
Dr Helmut Grimm

Number of full-time employees:
32

Competent Authority
Ministry for the Environment, Nature and Agriculture, Kiel

Information Centres
Multimar Wattforum
Am Robbenberg
D-25832 Tönning
Tel. +49 4861/9620-0
[daily 01.04. to 31.10.: 9.00 a.m. - 7.00 p.m.;
1.11. to 31.03.: 10.00 a.m. - 6.00 p.m.]

20 more information centres run by the NationalparkService gGmbH and nature conservation associations on all of the major islands and in many places on the mainland

Information Material
NationalparkService
Tel.: +49 4861/616-70
e-mail: service@nationalparkservice.de

Homepage
www.wattenmeer-nationalpark.de

Area
Total:
285,000 ha

Core area:
85,500 ha

Buffer zone:
6,400 ha

Transition area:
193,100 ha

Ownership (total area):
almost 100 % state

Geographical Location
Wadden Sea on the Schleswig-Holstein North Sea coast between Denmark in the north and the Elbe estuary in the south

Nature Space and Ecosystem
mudflats, salt marshes, dunes, beaches and sands, shallow water area of the North Sea (max. 20 m)

Population in the Biosphere Reserve
2 (permanent), 10 (summer)

Fulfilment of the Functions
Conservation:
national park in 1985; biosphere reserve in 1990; Ramsar area in 1991, Special Protection Area (SPA) according to EU Birds Directive and area of Community significance under Article 4, para. 2 Habitat Directive pursuant to Article 33 Federal Nature Conservation Act; biotope protection pursuant to Article 15a Land Nature Conservation Act

Development:
sustainable tourism development; diverse offers for holiday-makers by the nature conservation associations and the administration; development of detailed management regulations to ensure sustainable use of the area: fishing for shrimps (Crangon crangon) and mussels (Mytilus edulis), sheep grazing on the foreshore salt marshes

Logistic Support:
comprehensive monitoring programmes in agreement with the Netherlands, Denmark, Lower Saxony and Hamburg; regular basic and advanced training courses for multipliers (e.g. mudflat guides, personnel of nature conservation organisations); numerous projects for regional development in cooperation with Nationalpark Service gGmbH (incl. Multimar Wattforum) and the nature conservation associations

Regular Events
every year around 10,000 natural history excursions and lectures

Sponsoring Agency
none

Partnership
since 1992 partnership with the Russian Taimyrskiy Nature Reserve in northern Siberia

Special Features
area corresponds to that of the former National Park (National Park Act of 1985); an extension has currently been applied for; the inhabited islands are not part of the National Park

Where the Sea Floor Meets the Horizon

Hamburg Wadden Sea Biosphere Reserve

Federal Land
Hamburg (HH)

Year of UNESCO Recognition
1992

Administration
Nationalpark- und Biosphärenreservatsverwaltung
Hamburgisches Wattenmeer
Billstr. 84
D-20539 Hamburg
Tel.: +49 40/42845-3945
Fax: +49 40/42845-2579
e-mail: Klaus.Janke@bug.hamburg.de

Director:
Dr Klaus Janke

Number of full-time employees:
3

Competent Authority
Authority for Environment and Health
of the Free and Hanseatic City of Hamburg, Hamburg

Information Centre
Infozentrum Neuwerk
Information hotline: +49 4721/28594
from April 2004:

Nationalpark-Haus Neuwerk
27499 Insel Neuwerk
Tel.: +49 4721/395349
Fax: +49 4721/395866
Opening hours available from Nationalpark-Haus

Information Material
cf homepage

Homepage
www.wattenmeer-nationalpark.de
www.nationalpark-hamburgisches-wattenmeer.de

Area
Total
11,700 ha
Ownership:
99.77 % state, 0.23 % private

Core area
10,530 ha
Ownership:
100 % state

Buffer zone
1.170 (10 %)
Ownership:
97.7 % state, 2.3 % private

Transition area
none

Geographical Location
Wadden Sea/Elbe estuary

Nature Space and Ecosystem
Wadden Sea (including permanent tidal zone, tidal inlets, sandbanks, open mudflats, dunes, salt marshes and pastures)

Population in the Biosphere Reserve
approximately 40 (only Neuwerk island at approx. 3 km²)

Fulfilment of the Functions
Conservation:
protection and conservation of the natural dynamics in the Wadden Sea habitat; designated pursuant to Habitat/EC Bird Protection Directive

Development:
promoting extensive grazing and grassland farming to conserve the small farming structures and also promoting the reproductive success of the native meadow birds, i.e. Lapwing (Vanellus vanellus), Common Redshank (Tringa totanus), Sky Lark (Alauda arvensis), including monitoring

Logistic Support
environmental monitoring programme as integral component of Trilateral Wadden Sea Monitoring (DK/DE/NL) for environmental monitoring for Habitat/EC Bird Protection Directive

Regular Events
none

Sponsoring Agency
none

Partnerships
neighbouring Wadden Sea biosphere reserves

Special Features
Biosphere Reserve covers the identical area as the National Park of the same name.

Where the Sea Floor Meets the Horizon

Lower Saxon Wadden Sea Biosphere Reserve

Federal Land
Lower Saxony (NI)

Year of UNESCO Recognition
1993

Administration
Nationalparkverwaltung Niedersächsisches Wattenmeer
Virchowstr. 1
D-26382 Wilhelmshaven
Tel.: +49 4421/911-0
Fax: +49 4421/911-280
e-mail: dezernat04whv@br-we.niedersachsen.de

Director:
Irmgard Remmers

Number of full-time employees:
28 (of which 20 are full-time and 8 are part-time employees)

Competent Authority
Lower Saxon Environment Ministry, Hanover

Information Centres
Das Wattenmeerhaus
Südstrand 110b
D-26382 Wilhelmshaven
Tel.: +49 4421/9107-0

[April to October: daily 10.00 a.m. - 6.00 p.m.;

November: Tue-Sun 10.00 a.m. - 5.00 p.m.; 01.-24.12.: closed; 25.12. - 31.3.: Tue-Sun 10.00 a.m. - 5.00 p.m.]

another 14 National Park Houses and Centres, also on the islands (cf homepage)

Information Material:
cf homepage

Homepage
www.wattenmeer-nationalpark.de

Area
Total:
approximately 240,000 ha
Ownership: 99 % Land/Federal Republic

Core area:
approximately 130,700 ha
Ownership: 99 % Land/Federal Republic

Buffer zone
108,000 ha
Ownership: 98 % Land/Federal Republic

Transition area
approximately 2,000 ha
Ownership: 99 % Land

Geographical Location
North Sea coast of Lower Saxony from the base of the sea dike on the mainland up to a line on the sea side of the East Friesian Islands and the Platen (sands) and sandbanks in the Elbe-Weser estuary triangle; westernmost limit Aussenems (Ems) near Borkum, easternmost Kugelbake on the Elbe estuary near Cuxhaven

Population in the Biosphere Reserve
1

Fulfilment of the Functions
Conservation:
almost the entire core area and buffer zone and parts of the transition area protected as national park; establishment of large protected areas for seals (Pinnipedia) and birds

Development:
sustainable tourism development (visitor information and guidance); equalisation payments for farmers for extensive land use; care of land with local farmers within the context of contractual nature conservation; cooperation with the coastal protection administration, e.g. in drawing up foreshore management plans

Logistic Support:
conducting ecosystem research in Lower Saxon Wadden Sea; mudflat and visitor guide further training; educational and public relations work via the National Park Houses and Centres; sponsoring projects to improve and develop more sustainable catching methods in fisheries via the Wattenmeerstiftung (Wadden Sea Foundation); provision of materials for educational, public relations and information work

Regular Events
cf homepage

Sponsoring Agency
"Die Muschel" - Verein der Förderer und Freunde des Nationalparks Niedersächsisches Wattenmeer e. V.
Schleusenstr. 1
D-26382 Wilhelmshaven
Tel.: + 49 4421/944100

Partnerships
neighbouring biosphere reserves in the Wadden Sea

Special Features
Biosphere Reserve largely covers the identical area as the National Park of the same name; decision by the Lower Saxon Land Parliament to apply for recognition of the area as a UNESCO World Heritage Site

6. BIOSPHERE RESERVES IN GERMANY: AN OVERVIEW

Resting Place for Migrating Birds

Schaalsee Biosphere Reserve

Federal Land
Mecklenburg-Vorpommern (MV)

Year of UNESCO Recognition
2000

Administration
Amt für das Biosphärenreservat Schaalsee
Wittenburger Chaussee 13
D-19246 Zarretin
Tel.: +49 38851/302-0
Fax: +49 38851/320-20
e-mail: poststelle@schaalsee.mvnet.de

Director:
Klaus Jarmatz

Number of full-time employees:
34

Competent Authority
Mecklenburg-Vorpommern Environment Ministry and Mecklenburg-Vorpommern Ministry for Food, Agriculture, Forestry and Fisheries, Schwerin

Information Centre
PAHLHUUS Information Centre
(for address, cf. Administration)
[Mon-Fri 9.00 a.m. - 5.00 p.m.; Sat, Sun: 10.00 a.m. - 6.00 p.m.]

Information Material
Basic leaflet "Vielfalt erleben" [Experience Diversity]; "Den Schaalsee erleben" [Experience the Schaalsee]; "Natur und Kultur erleben 2003" [Experience Nature and Culture 2003]; "Porträt einer Landschaft" [Portrait of a Landscape]; "Die Marke für Ihr Wohlbefinden" [The Brand for Your Wellbeing]; "Offizielle Rad- und Wanderkarte Biosphärenreservat Schaalsee" [Official Schaalsee Biosphere Reserve Cycling and Hiking Map]; 3 x year information sheet "Biosphärenreservat Schaalsee aktuell" [Schaalsee Biosphere Reserve Now]; quarterly "Regionalmarke aktuell" [Regional Brand Now]

Homepage
www.schaalsee.de

Area
Total:
30,899 ha
Ownership: approximately 47 % state, 53 % private

Core area:
1,709 ha
Ownership: approximately 90 % state, 10 %: private

Buffer zone:
7,904 ha
Ownership: approximately 82 % state, 18 % private

Transition area:
21,286 ha
Ownership: approximately 30 % state, 70 % private

Geographical Location
Western Mecklenburg lake and hill region

Nature Space and Ecosystem
Bio-geographical province of Central and Eastern European forests

Population in the Biosphere Reserve
approximately 22,000 (approximately 71 per km²)

Fulfilment of the Functions
Conservation
various monitoring projects; moor, waterbody and small waterbody renaturalisation; species conservation projects, especially for the Fire-bellied Toad (Bombina bombina) and indicator species of the Schaalsee EC Bird Protection Directive; grassland extensification contracts; installing fish ladders; new forestation to buffer lakes to reduce nutrient inputs; woody planting; maintenance of oligotrophic sites

Development
Agenda 21 process; regional brand "Für Leib und Seele"; monthly Schaalsee Biosphere Market with produce from the region; "Theater im PAHLHUUS"; encouraging tourism by expanding the infrastructure (observation towers, footpath signposts, information boards)

Logistic Support
information centre in PAHLHUUS and exhibition in GRENZHUUS; intensive cooperation with schools and other educational facilities; "Biosphere Job Motor"; mentoring interns and undergraduates; specialist excursions; visitor mentoring; issuing own publications; visitor information

Regular Events
"TiP" Theater im PAHLHUUS; "Natur und Kultur erleben" [Experience Nature and Culture] events calendar; "Junior Rangers" working group

Sponsoring Agency
Förderverein Biosphäre Schaalsee e.V.
Wittenburger Chausse 13
D-19246 Zarrentin
Tel.: +49 38851/302-31

Partnerships
Schorfheide-Chorin BR; South-East Rügen BR; Colombian National Park administration; biosphere reserves currently being established in Iran and in the Baltic

Special features
location on the "Green Belt", the former border area between the two German states, with cultural landscape rich in species; plethora of various mosaic, interlinked small and large biotopes; numerous lakes and small waterbodies, moors, old beech forests, oligotrophic grassland and wetland meadows, but also highly productive agricultural land and old avenues

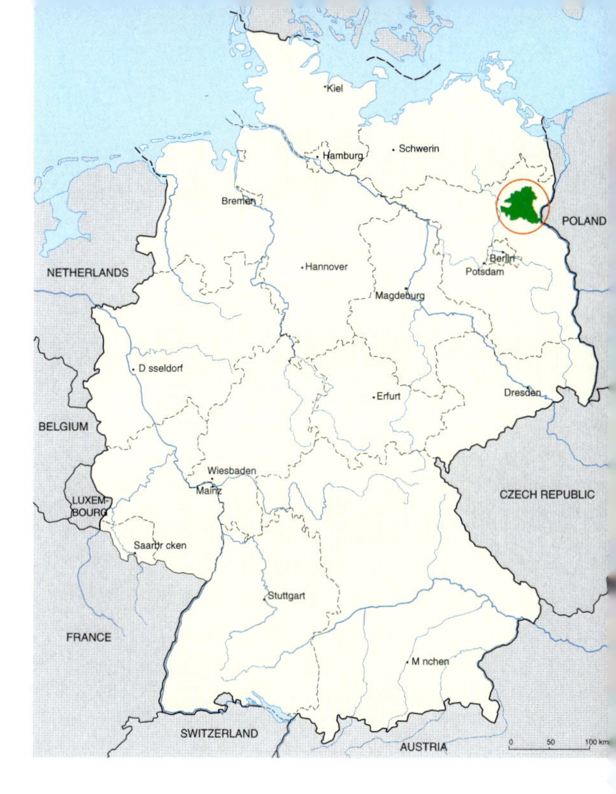

Land of Cranes with a Vast Horizon

Schorfheide-Chorin Biosphere Reserve

Federal Land
Brandenburg (BB)

Year of UNESCO Recognition
1990

Administration
Biosphärenreservatsverwaltung
Hoher Steinweg 5-6
D-16278 Angermünde
Tel.: +49 3331/3654-0
Fax: +49 3331/3654-10
e-mail: Monika.syhring@lags.brandenburg.de

Director:
Dr Eberhard Henne

Number of full-time employees:
18

Competent Authority
Ministry for Agriculture, Environmental Protection and Planning of the Land Brandenburg, Potsdam

Information Centres
Bürgerbüro des Biosphärenreservats Schorfheide-Chorin
Töpferstraße 1
D-16247 Joachimsthal
Tel. +49 33361/63380

"Blumberger Mühle" NABU Main Information Centre of the Schorfheide-Chorin BR
Tel. +49 3331/26040

Wildpark Schorfheide gGmbH
Prenzlauer Str. 16
D-16348 Groß Schönebeck
Tel. +49 33393/65855

For opening hours cf homepage

Homepage
www.schorfheide-chorin.de

Area
Total:
129,161 ha

Core area:
3,648 ha

Buffer zone:
24,650 ha

Transition area:
100,863 ha

Geographical Location
around 75 km north east of Berlin

Nature Space and Ecosystem
end moraine and basic moraine shaped by the Ice Age and sandy landscapes, forests, moors, lakes, open landscape

Population in the Biosphere Reserve
approximately 31,000 (approximately 25 per km^2)

Fulfilment of the Functions
Conservation:
renaturalisation projects; special species conservation programmes for the many animal and plant species at risk of extinction; programmes to conserve at risk crop plants (old grain, potato, vegetable and fruit varieties) by means of sponsoring agencies

Development:
development of sustainable forms of management with individual land users or companies; introduction of the Biosphere Reserve regional brand to build up regional economic cycles; agreement of forestry planning with the Biosphere Reserve maintenance and development plans, certification of the wood products; development of sustainable fisheries; further development of tourism

Logistic Support:
projects in the field of public relations and environment education with regional partners in the public relations working group of the Biosphere Reserve sponsoring agency; countless activities by the Nature Rangers, especially for children; citizens' office in Joachimstal, „Adebar" newspaper; coordination of research work with partners in the region; ecosystemic environmental monitoring

Regular Events
campaign days by the Biosphere Reserve; lectures and exhibitions in the citizens' office; guided tours by the Nature Rangers; meetings for children's groups; for more details cf homepage

Sponsoring Agency
Kulturlandschaftsverein Uckermark e. V.
Hoher Steinweg 5-6
D-16278 Angermünde
Tel.: +49 3331/298082

Partnerships
Sierra de las Nieves BR (Spain)
Issyk-Kul BR (Kyrgyzstan)

6. BIOSPHERE RESERVES IN GERMANY: AN OVERVIEW

World Culture on Wild Shores

Elbe River Landscape Biosphere Reserve

Federal Länder

Schleswig-Holstein (SH)
Mecklenburg-Vorpommern (MV)
Lower Saxony (NI)
Brandenburg (BB)
Saxony-Anhalt (ST)

Year of UNESCO Recognition

overall recognition: 1998

Administrations

SH
Landesamt für Natur und Umwelt Schleswig-Holstein (SH)
Hamburger Chaussee 25
D-24220 Flintbek
Tel.: + 49 4347/704-347
e-mail: jgemperl@lanu.landsh.de

MV
Naturparkverwaltung Mecklenburgisches Elbetal
Am Elbberg 20
D-19258 Boizenburg
Tel: +49 38847/624840
e-mail: naturparkverwaltung@elbetal-mv.de

NI
Bezirksregierung Lüneburg
Biosphärenreservatsverwaltung Niedersächsische Elbtalaue (NI)
Am Markt 1
D-29456 Hitzacker
Tel.: +49 5862/96730
e-mail: elbtalaue@br-lg.niedersachsen.de

BB
Biosphärenreservat Flusslandschaft Elbe - Brandenburg
Neuhausstr. 9
D-19322 Rühstädt
Tel.: +49 38791/980-0
e-mail: br-flusslandschaft-elbe@lags.brandenburg.de

ST
Biosphärenreservatsverwaltung Flusslandschaft Mittlere Elbe
Kapenmühle, PF 13 82
D-06813 Dessau
Tel.: +49 34904/421-0
e-mail: bioresme@t-online.de

Directors:

SH: Jürgen Gemperlein
MV: Eckhard Steffen
NI: Elvyra Kehbein
BB: Dr Frank Neuschulz
ST: Guido Puhlmann

Number of full-time employees

110 (SH: 1, MV: 11, NI: 10, BB: 19, ST: 69)

Competent Authorities

SH: Ministry for Environment, Nature Conservation and Agriculture of the Land Schleswig-Holstein, Kiel
MV: Environment Ministry Mecklenburg-Vorpommern, Schwerin
NI: Lower Saxon Environment Ministry, Hannover
BB: Ministry for Agriculture, Environmental Protection and Planning of the Land Brandenburg, Potsdam
ST: Ministry for Agriculture and Environment of the Land Saxony-Anhalt, Magdeburg

Information Centres

SH, MV
none

NI
Elbschloss Bleckede
Schlossstr. 10
D-21354 Bleckede
Tel.: +49 5852/95 14 0
e-mail: info@elbschloss-bleckede.de
[April-October: Tue-Sun 10.00 a.m. - 6.00 p.m.; November-March: Wed-Sun 10.00 a.m. - 5.00 p.m.]

BB
Rühstädt Visitor Centre
Neuhausstr. 9
D-19322 Rühstädt
Tel.: +49 38791/98022
e-mail: ruehstaedt_naturwacht@gmx.de

Burg Lenzen Visitor Centre
Burgstr. 3
D-19309 Lenzen
Tel.: +49 38792/1221
e-mail: Burg-lenzen@t-online.de
[daily 10.00 a.m. - 6.00 p.m.]

ST
Auenhaus Information Centre
Am Kapenschlösschen 3
D-06785 Oranienbaum
Tel: +49 34904/406-31
e-mail: bioresme-info@t-online.de
[May-October: Mon-Fri 10.00 a.m. - 5.00 p.m.; Sat, Sun and public holidays: 11.00 a.m. - 5.00 p.m.; November-April: Tue-Fri 10.00 a.m. - 4.00 p.m.; Sat, Sun and public holidays: 1.00 - 6.00 p.m.]

Information Material

e.g.: Nature Tourist Guide: "Elbtalaue - Landschaft am großen Strom" [Elbe Valley Meadows - Landscape Along a Major River]; Flyer "Weltkultur an wilden Ufern" [World Culture on Wild Shores]

Homepages

SH: none
MV: www.elbetal-mv.de
NI: www.bezirksregierung-lueneburg.de
BB: www.grossschutzgebiete.brandenburg.de
ST: www.BiosphaerenreservatMittlereElbe.de

Area

Total:

approximately 350,370 ha
no information is currently available on zoning and ownership conditions

Geographical Location

semi-natural river landscape with rivers along the Central Elbe

Nature Space and Ecosystem

Central Elbe Plain; Northern German Plain: river course with shorelines, flood plains (wet grassland, herbaceous land, alluvial woodland, dead arms, valley sand and inland dunes), woodland on slopes as well as neighbouring deciduous/mixed woodland, extensively used grassland

Population in the Biosphere Reserve

SH: 0,
MV: approximately 25,000 (63 per km²)
NI: 20,000 (35 per km²)
BB: 12,500 (23 per km²)
ST: no information

Fulfilment of the Functions

Conservation:

protection/development of the Elbe meadows with their traditional natural and cultural landscape, their uniqueness and beauty, conservation of the foundations of life of the diverse animal and plant communities,
protection and development of the complete biotope sequence in the vicinity of the Elbe bank and shallow water, protection/development of the Dessau/Wörlitz cultural landscape

Development:

project "Guiding Principles of Nature Conservation and their Implementation with Agriculture - Objectives, Instruments and Costs of an Environmentally-Friendly and Sustainable Agriculture in the Lower Saxon Elbe Valley"; project "Alluvial Meadow Regeneration by Setting Back Dykes" (BB) within the framework of the Federal Ministry for Education and research alliance "Elbe Ecology"; various EU LIFE Projects to develop Natura 2000, Project "Biosphere Reserve Shop Window" to set up information facilities and landscape-based area and project information; regional brand; LEADER plus project; Biosphere Reserve tours; EU LIFE project in Klieken; EU INTERREG IIIb; Wetlands II Project; major nature conservation project "Central Elbe" run by the BfN with the WWF

Logistic Support:

setting up nature teaching paths and nature experience route, cooperation with schools in the region; placement, undergraduate and doctorate theses on various subjects; setting up a documentation system; monitoring for Natura 2000 areas, ecosystemic environmental monitoring

Regular Events

SH: none

MV: comprehensive annual programme, annual "Naturpark-Festival", annual newsletter "Naturpark-News"

NI: series of lectures "Elbe Evenings"

BB: comprehensive annual programme, annual conferences, junior ranger programmes, summer campaigns, Biosphere Reserve tours, "Long Nights in the Rühstädt Visitor Centre"

ST: e.g. specialist conferences, open day, schoolchildren's events (kick-start), "Gartenreich" Day

Sponsoring Agencies

SH: not in place

MV: Förderverein Naturpark Mecklenburgisches Elbetal e. V.
Am Elbberg 20
D-19258 Boizenburg
Tel.: +49 38847/54755

NI
Förderverein Naturschutz Elbetal e. V.;
Förderverbund Elbtalaue e. V.
Rohrberg 2
D-29439 Lüchow
Tel.: +49 5841/973655

BB
Förderverein Biosphärenreservat Flusslandschaft Elbe - Brandenburg e. V.
D-19309 Lenzen
Seestr. 18
Tel.: +49 38792/1701

ST
Förder- und Landschaftspflegeverein "Biosphärenreservat Mittlere Elbe" e. V.
Albrechtstr. 128
D-06844 Dessau
Tel.: +49 340/2206141

Partnerships

BB
member of the EUROSITE Network

Special Features

Biosphere Reserve covering five Federal Länder along the Elbe; transboundary-Land coordination centre in Havelberg, biggest contiguous alluvial meadows in Central Europe

Pick Up Your Pickles in a Gondola

Spree Forest Biosphere Reserve

Federal Land
Brandenburg (BB)

Year of UNESCO Recognition
1991

Administration
Biosphärenreservat Spreewald
Schulstr. 9 · D-03222 Lübbenau
Tel.:+ 49 3542/8921-0 · Fax: +49 3542/8921-40
e-mail: br-spreewald@lags.brandenburg.de

Director: *Eugen Nowak*

Number of full-time employees: *19*

Competent Authority
Ministry for Environmental Protection,
Agriculture and Planning of the Land Brandenburg, Potsdam

Information Centres
Haus für Mensch und Natur Lübbenau
Schulstr. 9 · 03222 Lübbenau
Tel.: 03542/8921-0

Burg Information Centre
Byhleguhrer Str. 17 · D-03096 Burg
Tel.: +49 35603/691-0

Schlepzig Information Centre
Dorfstr. 52 · D-15910 Schlepzig
Tel.: +49 35472/648-98

[from April to September daily 10.00 a.m. - 5.00 p.m. and as arranged]

Information Material
"Adebar" (Newspaper for the population in the Biosphere Reserve); leaflets on special subjects: agriculture, forest, canoeing, water hiking maps with tour recommendations, animal and plant species, environmental education and information offers, Spree Forest Report (popular scientific portrayal of results of work, research, etc.); CD-ROM

Homepage
www.spreewald.de
www.grossschutzgebiete.brandenburg.de/br_spree/

Area
Total: *47,509 ha*
Ownership: 14 % Land, 6 % BVVG, 53 % private, 9 % community property, 18 % other

Core area: *974 ha*
Ownership: 36 % Land, 28 % BVVG, 11 % private, 1 % Federal Republic, 1 % local authorities, 16 % other, 7 % unknown

Buffer zone: *9,334 ha*
Ownership: 22 % Land, 10 % BVVG, 45 % private, 2 % local authorities, 20 % other, 1 % unknown

Transition area: *37,201 ha*
Ownership: 11 % Land, 4 % BVVG, 56 % private, 12 % community property, 17 % other

Geographical Location
100 km south east of Berlin

Nature Space and Ecosystem
largely semi-natural alluvial meadow landscape with approximately 1,550 km flowing waters

Population in the Biosphere Reserve
49,700, of which 30,500 in the two towns of Lübbenau and Lübben, 19,200 in the rural area (99 per km²; 38 per km² in the rural area)

Fulfilment of the Functions
Conservation:
for approximately 2,000 ha grassland extensification programmes (approximately 4,000 ha wetland meadows by means of contractual nature conservation via Spree Forest BR; Spree Forest Water Edge Project (at the planning stage); conservation programmes for otters (Lutra lutra) and White Stork (Ciconia ciconia) in cooperation with NGOs and sponsors

Development:
special Programme for the villages Lehde and Leipe - conservation of the historically developed, small-scale cultural landscape including the traditional farming and craft skills; Spree Forest Meadow Programme - Special Programme in the context of the EU co-financed cultural landscape programme; Spree Forest regional brand; Spree Forest soil ordinance programme; conservation programme of old cultural varieties

Logistic Support:
ecosystemic environmental monitoring on 30 permanent monitoring sites or stretches of flowing water, collection of up to 295 different parameters; development and implementation of techniques to protect the soil on lowland moor soils and in forestry and agriculture; setting up a system of reference farms in the agricultural sector; introduction of the „REPRO Farm Balancing and Indicator Model" on farms; Spree Forest Nature Rangers

Regular Events
natural history walks; cycle tours and events on special subjects; public relations days in the Nature Experience Clock; guided tours in the Burg herb garden; high points such as Day of Parks, children's festivals

Sponsoring Agencies:
Förderverein für Naturschutz im Spreewald (FÖNAS)
Schulstr. 9 · D-03222 Lübbenau
Tel.: +49 3542/8921-0

Zweckverband Gewässerrandstreifenprojekt Spreewald
Ehm-Welk-Str. 15 · D-03222 Lübbenau
Tel.: +49 3542/872817

Freundeskreis des Gewürzpflanzengartens Burg e. V.
Byhleguhrer Str. 17 · D-03096 Burg
Tel.: 035603/69124

Carpus e. V. (for partnership with Palawan BR/Philippines)
Nordweg 7 · D-03096 Burg/Spreewald
Tel.: +49 35603/69123

Partnerships
Palawan BR (Philippines), contract until 2007
Staatsbosbeheer Nationalpark „De Weerribben" (Netherlands), contract until 2004

Carp Ponds on the Heath

Upper Lausitz Heath and Pond Landscape Biosphere Reserve

Federal Land
Saxony (SN)

Year of UNESCO Recognition
1996

Administration
Verwaltung des BR Oberlausitzer Heide- und Teichlandschaft
Am Sportplatz 231
D-02906 Mücka
Tel.: +49 35893/506-40
Fax: +49 35893/506-50
e-mail: poststelle@brv.smul.sachsen.de

Director:
Peter Heyne

Number of full-time employees:
14

Competent Authority
Saxon State Ministry for Environment and Agriculture, Dresden

Information Centre
currently only at the administration (Verwaltung)
[Mon-Thu 8.00 a.m. - 4.00 p.m., Fri 8.00 a.m. - 2.00 p.m.]
Information centre at the planning stage

Information Material
countless material, cf homepage

Homepage
www.biosphaerenreservat-oberlausitz.de

Area
Total:
30,102 ha
Ownership: 15 % Land, 10 % Federal Republic, 5 % local authorities, 6 % nature conservation association, 64 % private

Core area:
1,124 ha
Ownership: 25 % Land, 25 % Federal Republic, 30 % nature conservation association, 20 % private (in the process of being purchased)

Buffer zone:
12,015 ha
Ownership: 35 % Land, 25 % Federal Republic, 5 % local authorities, 3 % nature conservation association, 32 % private

Transition area:
16,963 ha
Ownership: 5 % Land, 3 % Federal Republic, 6 % local authorities, 6 % nature conservation association, 80 % private

Geographical Location
eastern Saxony

Nature Space and Ecosystem
Upper Lausitz heath and pond area

Population in the Biosphere Reserve
12,800 (43 per km²)

Fulfilment of the Functions
Conservation:
numerous species and biotope conservation measures for White-tailed Eagle (Haliaeetus albicilla), Eurasian Otter (Lutra lutra), bats (Chiroptera); wetland meadow maintenance; ecological consistency of the Spree; conservation of old crop plants and arable weeds and much more); comprehensive protective measures (emphasis on fisheries and grassland maintenance); close cooperation with managers and authorities with regard to support for species and biotope conservation measures; maintenance and development plans for specific areas

Development:
encouraging environmentally friendly management; conception for sustainable development and regeneration; support for regional marketing; conception for settlement development, concepts on tourism and traffic trends; Biosphere Reserve logo as seal of quality for products and services; nature markets

Logistic support:
incentives and participation in demonstration projects; environmental education for children and adults; comprehensive environmental monitoring conception; cooperation contracts with scientific facilities; regular analysis of water and precipitation data; climate station

Regular Events
project days in all schools; numerous nature guided tours and lectures; two specialist colloquia per year; practical seminars for the public; four nature markets per year; numerous environmental education events

Sponsoring Agency
Förderverein für die Natur der
Oberlausitzer Heide- und Teichlandschaft
An der Post 2
D-02906 Kreba
Tel.: +49 35893/50266
e-mail: Foerderverein-oberlausitz@t-online.de

Partnerships
Spree Forest and Elbe River Landscape BR
Trebon BR (Czech Republic)

Special Features
bilingual region (German and Sorbian)

6. BIOSPHERE RESERVES IN GERMANY: AN OVERVIEW

Peaceful Forests and Murmuring Brooks

Vessertal-Thuringian Forest Biosphere Reserve

Federal Land
Thuringia (TH)

Year of UNESCO Recognition
1979 (1st extension 1986, 2nd extension 1990)

Administration
Biosphärenreservat Vessertal
Verwaltung
Waldstr. 1 · D-98711 Schmiedefeld am Rennsteig
Tel.: +49 36782/666-0 · Fax: +49 36782/666-29
e-mail: poststelle.vessertal@br-np.thueringen.de

Director:
Johannes Treß

Number of full-time employees:
13

Competent Authority
Thuringian Ministry for Agriculture,
Nature Conservation and Environment, Erfurt

Information Centre
Information and Education Centre
Biosphärenreservat Vessertal
Nordstr. 96 · D-98711 Frauenwald
Tel.: +49 36782/62947
e-mail: br-vessertal@t-online.de
[Mon-Fri 9.00 a.m. - 4.00 p.m. on request;
on other days by appointment]

Information Material
Brochures, leaflets (general, about important habitats, for children); series of posters; hiking map „Vessertal Biosphere Reserve"; tear-off pads with recommendations for local walks

Homepage
www.biosphaerenreservat-vessertal.de

Area
Total: *17,098 ha*
Ownership: 88 % Land, 12 % private and corporations

Core area: *437 ha*
Ownership: 97 % Land, 3 % private and corporations

Buffer zone: *2,024 ha*
Ownership: 90 % Land, 10 % private and corporations

Transition area: *14,637 ha*
Ownership: 87 % Land, 13 % private and corporations

Geographical Location
Central Thuringian Forest between Ilmenau, Schleusingen and Suhl

Nature Space and Ecosystem
medium-range mountains, Central Thuringian Forest, forests in temperate Europe, e.g. woodrush beech forest (Luzulo-Fagetum); semi-natural grassland, e.g. milkwort mat grassland (Polygalo-Nardetum); wood cranesbill yellow oat grass meadow (Geranio sylvatici-Trisetetum); upland moors; semi-natural flowing waters in mountainous areas

Population in the Biosphere Reserve
approximately 4,200 (approximately 25 per km²)

Fulfilment of the Functions
Conservation:
conservation of semi-natural forests; conversion of pure spruce stands to semi-natural mixed mountain woods; maintenance and development of semi-natural woodland streams and sparse woodland habitats; gene conservation plantation for Silver Fir (Albies alba); encouraging extensive grassland use by mowing and grazing; conservation and renaturalisaiton of moors; conservation of semi-natural mountain streams and renaturalisation of developed stretches

Development:
sustainable tourism development; sustainable forestry according to the principles of semi-natural silviculture; implementation of the maintenance and development plan - specialist part on forest - for selected areas; encouraging agriculture in medium-range mountains by advising and mentoring farmers and providing support in applications for funding with the objective of conserving the open landscape

Logistic support:
environmental education and information; specialist excursions; natural history walks; environmental education programmes and project days for school classes; information and education centre; information garden; exhibitions; various teaching paths and information stand; development of environmental quality goals with cause-effect-hypotheses; operating environmental monitoring facilities by the German Meteorological Service, etc.; compiling a lead and indicator species concept; keeping a bibliography; stocking profile analysis in the core area; initial forestry stocktaking in the „Vessertal" and „Marktal and Morast" natural forest plots; research project on biomanipulation in the Schönbrunn dam; provision of data for Biosphere Reserve Integrated Monitoring (BRIM) and for various meta-databases on the internet

Regular Events
thematic guided tours for visitor groups; holiday programmes; project days for school classes; slide presentations; European Day of Parks every May

Sponsoring Agency
Förderverein Biosphärenreservat Vessertal-Thüringer Wald e. V.
Nordstr. 96 · D-98711 Frauenwald
Tel.: +49 36782/62947
e-mail: br-vessertal@t-online.de

Special Features
BR administration also responsible for performing the state tasks in the Thuringian Forest Nature Park in the field of nature conservation and landscape management; the "Rennsteig", the ridge path in the Thuringian Forest, 168 km long and one of the most famous upland walking routes in German crosses practically through the middle of the Biosphere Reserve.

Land of Open Vistas

Rhön Biosphere Reserve

Federal Länder

Hesse (HE)
Thuringia (TH)
Bavaria (BY)

Year of UNESCO Recognition

1991

Administrations

HE
Landrat des Kreises Fulda
Abt. Amt für ländlichen Raum Hessische Verwaltungsstelle
Biosphärenreservat Rhön
Groenhoff-Haus / Wasserkuppe
D-36129 Gersfeld
Tel.:+49 6654/9612-0
e-mail: vwst@biosphaerenreservat-rhoen.de

TH
Biosphärenreservat Rhön Verwaltung Thüringen
Mittelsdorfer Str. 23
D-98634 Kaltensundheim
Tel.: +49 36946/382-0
e-mail: poststelle.rhoen@br-np.thueringen.de

BY
Regierung von Unterfranken
Bayerische Verwaltungsstelle Biosphärenreservat Rhön
Oberwaldbehrunger Straße 4
D-97656 Oberelsbach
Tel.: +49 9774/9102-0
e-mail: postmaster@brrhoenbayern.de

Directors:
HE: Heinrich Heß
TH: Karl-Friedrich Abe
BY: Michael Geier

Number of full-time employees:
19.5 (HE: 6, TH: 8, BY: 5.5)

Competent Authorities

HE: Hessian Ministry for Environment, Land Development and Consumer Protection, Wiesbaden
TH: Thuringian Ministry for Agriculture, Nature Conservation and Environment, Erfurt
BY: Bavarian State Ministry of the Environment, Public Health and Consumer Protection, Munich

Information centres

HE
Informationsstelle Biosphärenreservat Rhön (HE)
Groenhoff-Haus - Wasserkuppe, D-36139 Gersfeld
Tel.: +49 6654/96120
[Mon-Fri 7.30 a.m. - 4.00 p.m., Sat, Sun 10.00 a.m. - 4.00 p.m.]

Landschaftsinformationszentrum Rasdorf (HE)
Am Anger 32
D-36169 Rasdorf
Tel.: +49 6651/9601-0
[Opening hours by arrangement]

TH
Haus auf der Grenze/Point Alpha
Tel.: +496651/919030
[Nov. to March: daily 10.00 a.m. - 5.00 p.m.;
April to Oct.: daily 9.00 a.m. - 6.00 p.m.]

Propstei Zella/Rhön
Tel.: +49 36964/93510
[Tue-Fri 10.00 a.m. - 5.00 p.m.; Sat, Sun 1.00 - 5.00 p.m.]

BY
Info-Zentrum „Haus der Schwarzen Berge" (BY)
Rhönstr. 97
D-97772 Wildflecken-Oberbach
Tel.: +49 9749/9122-0
[April-October: Tue-Fri 10.00 a.m. - 6.00 p.m.;
Sat, Sun 10.00 a.m. - 5.00 p.m.]

Info-Zentrum „Haus der Langen Rhön" (BY)
Unterelsbacher Str. 4
D-97656 Oberelsbach
Tel.: +49 9774/910260
[April-October: Mon-Fri 9.00 a.m. - 5.00 p.m.;
Sat, Sun 10.00 a.m. - 5.00 p.m.; Tue closed]

Information Material

"The Rhön Biosphere Reserve - Always an Experience"; annual programmes of the information centres; notifications from the Biosphere Reserve; brochures; leaflets; slide shows; video films

Homepage

www.biosphaerenreservat-rhoen.de

Area

Total:
184,939 ha (of which HE: 63,564 ha, TH: 48,571 ha, BY: 67,102 ha)

Core area:
4,199 ha
Ownership: 83 % private, 1 % local authority, 16 % state

Buffer zone:
67,483 ha

Transition area:
107,557 ha
Ownership (buffer zones and transition areas):
information not yet collected
(5,700 ha of the Bavarian section have not yet been zoned)

Geographical Location

meeting point of three Länder between Hesse, Thuringia and Bavaria

Nature Space and Ecosystem

around central basalt highlands peaks of single, forested cone-shaped mountain-tops, transition to the Swabian-Franconian Cuesta Region; heights from approximately 230 to 950 m above sea level; agricultural use with the emphasis on grassland management in the higher positions, connected with hedges and forests; woodland proportion approximately 42 %. The forests are semi-natural and species-rich mountainous broadleaf forests around the upper new red sandstone and the upper shell limestone and primarily arable, sometimes even in the upper shell limestone; dry calcareous grassland on the steep slopes of the lower shell limestone as a result of sheep and cattle grazing, especially in TH.

Population in the Biosphere Reserve

135,618 (total: 79 per km²)
HE: 48,858
TH: 39,294
BY: 47,466

Fulfilment of the Functions

Conservation:

purchase and maintenance of land for the conservation and development of precious natural spaces; large-scale projects „Hohe Rhön/Lange Rhön" in Bavaria, „Thüringer Rhönhutungen" to conserve the chalk oligotrophic grasslands, further conservation measures through various funding programmes; model project „Landschaftspflege durch Großbetriebe" on landscape management at the biggest working farm in the Biosphere Reserve

Development:

focus of the development activities: setting up regional marketing across the Länder boundaries, including establishing cooperations between agriculture, the manufacturing industry and commerce and in cooperation with numerous partners from authorities and associations, e.g. marketing of products from traditional orchards, Rhön lamb products, beef, brown trout; including coupling with services in the field of leisure/tourism

Logistic support:

environmental education and public relations: Interface in the multifunctional information centres with a broad range of programmes; basic elements for recording environmental data through pilot project „Ecosystemic Environmental Monitoring"; setting up a GIS; research projects, e.g. R&D project „Monitoring Success in Major Nature Conservation Projects", T&D project „Conflict Solutions between Sport and Nature Conservation using the example of the Hohe Rhön", Federal Ministry of Education and Research project for large-scale stochastic grazing, research projects to monitor neophytes, habitat reconstruction experiments to convert intensive grassland into semi-natural meadows; systematic research of the forests in the core area in Bavaria, Hesse and Thuringia; Rhön Regional Working Group (ARGE) on cooperation across the Länder boundaries by district councillors, authorities, organisations, as part of the ARGE, including creating a „Rhön umbrella brand"

Regular Events

presentation of research results; excursions; events organised by the information centres; slide presentations; video films; bird population censuses; shepherds' assemblies; annual star-gazing walk with schoolchildren; exhibitions; presentations at trade fairs; Day of Species Diversity; scientific symposia; workshops with wood carvers, etc.

Sponsoring Agencies

HE
Verein Natur und Lebensraum Rhön e. V.
Groenhoff-Haus Wasserkuppe
D-36129 Gersfeld
Tel.: +49 6654/9612-0

BY
Verein Naturpark und
Biosphärenreservat Bayerische
Rhön e. V.
Oberwaldbehrunger Str. 4
D-97656 Oberelsbach
Tel.: +49 9774/910250
e-mail: info@brrhoenbayern.de

Partnerships

HE: Parrikkala region (Finland);
BY: cooperation with the Limousin region (France)

Special Features

Transboundary Biosphere Reserve across 3 Länder: organisation of cooperation between the administrative offices by the Administrative Agreement concerning the Establishment, Development and Administration of the Rhön BR (1.12.2002): Lead responsibility of one administrative office for three years, involving coordination of the projects concerning more than one Land

The Marriage of Forest and Vineyard

Palatinate Forest(D) - Northern Vosges (F) Transboundary Biosphere Reserve German Part

Land (in Germany)

Rhineland-Palatinate (RP)

Year of UNESCO Recognition

national: 1993
transboundary: 1998

Administration

Managing body for the German part:
Naturpark Pfälzerwald e. V.(NGO)
Franz-Hartmann-Str. 9 · D-67466 Lambrecht
Tel +49 6325/95520 · Fax +49 6325/955219
e-mail: info@pfaelzerwald.de

Director:

Werner F. Dexheimer

Number of full-time employees:

5

Competent Authority

Rhineland-Palatinate Ministry of Environment and Forestry, Mainz

Information Centres

Pfalzmuseum für Naturkunde
Hermann-Schäfer-Str. 17 · D-67098 Bad Dürkheim
Tel.: +49 6322/94130
[Tue-Sun 10.00 a.m. - 5.00 p.m.]

Biosphärenhaus Pfälzerwald/Nordvogesen
Am Königsbruch 1 · D-66996 Fischbach bei Dahn
Tel.: +49 6393/92100
Internet: www.biosphaerenhaus.de
[Tue-Sun 9.30 a.m. - 5.00 p.m.]

Information Material

Information brochures; leaflets, posters; scientific yearbook; tourist map; newspaper inserts; documentations; route guides

Homepage

www.pfaelzerwald.de (national)
www.biosphere-pfaelzerwald-vosges.org (transboundary, trilingual)

Area

Total:
177.842 ha (German part only)
Ownership: 70 % Land, 20 % municipal and local communities, 10 % private

Core area: *3,739 ha*
Ownership: 94 % Land, 6 % municipal and local communities

Buffer zone: *49,261 ha*
Ownership: 76 % Land, 24 % municipal and local communities

Transition area: *124,842 ha*
Ownership: approximately 70 % Land, 20 % municipal and local communities, 10 % private

Geographical Location

south-west of the Federal Republic of Germany, in the south of Rhineland-Palatinate; French part situated in the regions of northern Alsace/eastern Lorraine

Nature Space and Ecosystems

Palatinate Forest/Wine Route (D) and Northern Vosges (F): triassic coloured sandstone low mountain range(highlands); large-scale, close-to-nature terrestrial ecosystems (primarily: dense, deciduous, low mountain range forests of the temperate zone, also: grasslands capable of supporting trees, oligotrophic meadows, dwarf- heathlands and extensively managed cash-crop stands); small-range, close-to-nature, semi-terrestrial ecosystems (e.g.: bogs, mires, mire-like swamps); wide-ranging and small-scale, close-to-nature freshwater-ecosystems (e.g.: stagnant and running waters); primarily in the marginal area of the Biosphere Reserve: urban-industrial and agro-industrial ecosystems

Population in the Biosphere Reserve

approximately 160.000 (approximately 90 per km^2)

Fulfilment of the Functions

Conservation:
GIS-based management-plan for grazing to keep the valleys as open land; initiative „Pro Luchs" to protect the lynx (Lynx lynx); protection of the Wild Cat (Felis sylvestris), the Peregrine Falcon (Falco peregrinus), various bat species (Myotis spec.).

Development:
European Charter for Sustainable Tourism; marketing of regional produce; German-French rural markets ; „Biosphere Reserve partnership"-initiative for regional small-scale enterprises

Logistic Support:
use of wood for energy and construction purposes; wild-game marketing initiative; specialist working group on environmental education in order to interconnect local players and initiatives

Regular Events

'Wasgau discussion-forum'; lectures; presentations; guided tours

Sponsoring Agency

aimed at

Partnership

aimed at

Special Features

first Transboundary Biosphere Reserve (Germany/France) in the European Union
Address of the French managing body:
SYCOPARC
Maison du Parc - B.P. 24
F-67290 La Petite-Pierre
Tel.: +33 38801/4959, Fax: -60
e-mail: contact@parc-vosges-nord.fr
managing director: Marc Hoffsess

Boundless Forest Wilderness

Bavarian Forest Biosphere Reserve

Federal Land
Bavaria (BY)

Year of UNESCO Recognition
1981

Administration
Nationalpark- und Biosphärenreservatsverwaltung
Freyunger Str. 2
D-94481 Grafenau
Tel.: +49 8552/9600-0
Fax: +49 8552/9600-100
Email:poststelle@fonpv-bay.bayern.de

Director:
Karl Friedrich Sinner

Number of full-time employees:
193

Competent Authority
Bavarian State Ministry of the Environment, Public Health and Consumer Protection, Munich

Information Centres
Hans-Eisenmann-Haus
Böhmstr. 35
D-94556 Neuschönau
Tel.: +49 8558/96150
[daily 9.00 a.m. - 5.00 p.m. (winter: to 4.00 p.m.)]
Haus zur Wildnis (under construction)

Information Material
National Park educational and information offer (cf. home page)

Homepage
www.nationalpark-bayerischer-wald.de

Area
Total:
13,329 ha

Core area:
10,224 ha

Buffer zone:
3,105 ha

Transition area:
0

Ownership (total):
99 % Land, 1 % private and municipalities

Geographical Location
mountain range between „Mt Rachel"(1,453 m) and „ Mt Lusen" (1,373 m) along the German-Czech border in the Freyung-Grafenau district (south east Bavaria)

Nature Space and Ecosystem
inner Bavarian Forest (with uplands between 700 and 1,453 m); 99 % forested. Valley spruce forests, mixed mountain forests (spruce-beech-fir; especially luzolo-fagetum) on slopes, mountain spruce forests (soldanello-piceetum) in highland positions (over 1,150 m); special features: upland moors, block fields and an Ice Age lake (Lake Rachel).

Population
0

Fulfilment of the Functions
Conservation:
largest land area protected under nature conservation law in Germany; core area predominately for process conservation (biotic communities develop without human interventions according to their natural dynamics; renaturalisation and return to semi-natural state of the habitats more severely impaired by man (e.g. moors, stream courses); target species conservation measures for individual highly endangered animal and plant species, e.g. capercaillie (Tetrao urogallus), lynx (Lynx lynx), yew (Taxus baccata)

Development:
focus: development of sustainable, nature-friendly („gentle") tourism and environmentally friendly local public transport network („Igel" buses); support for nature-friendly regional agriculture and forestry (e.g. heating from wood chips, „Kiosk der Region", regional products in the wilderness camp)

Logistic Support:
„Jugendwaldheim" educational facility and „Wilderness Camp on Falkenstein": environmental education work for school classes with almost 10,000 pupils per year; over 3,000 educational events (guided tours, lectures, seminars) per year with over 40,000 people; so far in the BR approx. 600 research projects (focus on nature conservation); current development of an internet-based tourist GIS as part of the „high-tech offensive"

Regular Events
cf. homepage

Sponsoring Agency
Verein der Freunde des Ersten Deutschen Nationalparks Bayerischer Wald e. V.
Kröllstr. 5
D-94481 Grafenau
Tel.: +49 8552/9205-27
e-mail: info@nationalparkfreunde.de

Partnership
memorandum with Sumava National Park and Biosphere Reserve (Czech Republic)

Special Features
Biosphere Reserve covers the identical area as the National Park of the same name (former area); the biggest wilderness area in the whole of Central, Western and Southern Europe is being established in the core area.

Discover Nature in the Alps - and Savour Health

Berchtesgaden Biosphere Reserve

Federal Land

Bavaria (BY)

Year of UNESCO Recognition

1990

Administrations

Landratsamt Berchtesgadener Land
Salzburger Straße 64
D-83435 Bad Reichenhall
Tel.: +49 8651/773-521
Fax: +49 8651/773-599
e-mail: roland.beier@lra-bgl.de und

Nationalparkverwaltung Berchtesgaden
Doktorberg 6
D-83471 Berchtesgaden
Tel.: +49 8652/9686-0
Fax: +49 8652/9686-40
e-mail: m.vogel@nationalpark-berchtesgaden.de

Directors:
Roland Beier; Dr. Michael Vogel

Number of full-time employees:
70

Competent Authority

Bavarian State Ministry of the Environment, Public Health and Consumer Protection, Munich

Information Centres

Nationalpark-Haus Berchtesgaden
Franziskanerplatz 7
D-83471 Berchtesgaden
Tel.: +49 8652/64343
Fax: +49 8652/69434
[all year round daily 9.00 a.m. - 5.00 p.m.]

five more centres, cf homepage

Homepage

www.lra-bgl.de
and
www.nationalpark-berchtesgaden.de

Area

Total:
46,710 ha

Core area:
13,896 ha
Ownership: 100 % state

Buffer zone:
6,914 ha
Ownership: 100 % state

Transition area:
25,900 ha
Ownership: state, municipality, private

Geographical Location

south eastern Bavaria

Nature Space and Ecosystems

Berchtesgaden Alps, high mountains, forests

Population in the Biosphere Reserve

45,229 (97 per km²)

Fulfilment of the Functions

Conservation:
forest conversion, game management, forest-game management

Development:
National Park Plan

Logistic support:
ecosystem analysis; environmental monitoring using GIS; environmental education, i.e. visitor care, multiplier training, excursions, hikes

Regular Events

annual meeting of the sponsoring agency; working meetings of the National Park Advisory Council

Sponsoring Agency

Freunde des Nationalparks Berchtesgaden e. V.,
Doktorberg 6,
D-83471 Berchtesgaden
Tel.: +49 8652/9686-0
Fax: +49 8652/9686-40
e-mail: m.vogel@nationalpark-berchtesgaden.de

Partnership

none

Special Features

The core area and buffer zone cover exactly the identical area as the Berchtesgaden National Park.

Annex

7.

7.1 *National Catalogue of Criteria*

Extract from the "Criteria for Designation and Evaluation of UNESCO Biosphere Reserves in Germany", edited by the German National Committee for the UNESCO Man and the Biosphere (MAB) Programme:

Structural criteria

■ Representativeness

(1) The biosphere reserve must contain ecosystem complexes that, to date, are not sufficiently well-represented in biosphere reserves in Germany. (A)

■ Size (Area)

(2) The biosphere reserve should, as a rule, comprise at least 30,000 ha and should not be larger than 150,000 ha. Biosphere reserves that cross *Länder* boundaries may have a total area larger than this, if an appropriate level of administrative resources is provided. (A)

■ Zonation

(3) The biosphere reserve must be divided into core areas, buffer zones and transition areas. (A)

(4) The core area must take up at least 3 % of the total area. (A)

(5) The buffer zone should take up at least 10 % of the total area. (A)

(6) The core area and buffer zone, together, should account for at least 20 % of the total area. The core area should be surrounded by the buffer zone. (A)

(7) The transition area should take up at least 50 % of the total area; in marine areas, this requirement applies to the area on land. (A)

■ Legal Protection

(8) The biosphere reserve's protective purpose, and maintenance and development aims, both for the entire area and within the individual zones, must be protected by legal ordinances – or through *Land* and regional planning and programmes – and through development planning (Bauleitplanung) and landscape planning. On the whole, the majority of the area must be under legal protection. The protection status of existing core areas must not be downgraded. (B)

(9) The core area must be legally protected as a national park or nature reserve. (A)

(10) The buffer zone should be legally protected as a national park or nature reserve. Where this aim has not been achieved, the appropriate legal protection must be sought. (B)

(11) Areas within the transition area that are worthy of protection must be legally protected as designated protected areas and by development and landscape planning instruments. (B)

■ Administration and Organisation

(12) The biosphere reserve must have a capable administration – or such an administration must be established within three years. This administration must be appropriately staffed with technical and administrative personnel and appropriately equipped for its tasks. The application must include a commitment to provide the necessary funding. (A)

(13) The biosphere's administration must be organised as a part of the intermediate, higher, or highest nature conservation authority. The tasks of the biosphere reserve's administration, and relevant tasks of other existing administrations and other sponsors, must be clarified and co-ordinated from a work-sharing perspective. (B)

(14) The area must have a full-time administration. (B)

(15) The local population must be enabled to share in designing the biosphere reserve as its area for living, working and engaging in recreation. Proof must be supplied that suitable forms of citizens' participation are being practised. (B)

(16) Suitable non-profit or privately funded structures and organisation must be developed to handle tasks that can be partially or completely delegated. (B)

■ Planning

(17) A co-ordinated framework concept must be prepared within three years after the biosphere reserve has been designated by UNESCO. The application must contain a commitment to provide the necessary funding. (A)

(18) Maintenance and development plans should be prepared within five years, on the basis of the framework concept – at least for areas within the buffer zone and transition area that require particular protection or care. (B)

(19) The biosphere reserve's aims and the framework concept should be integrated, at the earliest possible time, within *Land* and regional planning and within landscape and development planning. (B)

(20) Aims for the biosphere reserve's protection, maintenance and development should be taken into account in updates of other technical planning. (B)

Functional criteria

■ Sustainable use and development

(21) Sustainable use and development of the biosphere reserve, and of the surrounding region, should be promoted, in all economic sectors, in keeping with regional and inter-regional possibilities and resources. Relevant administrative, planning and financial measures should be identified and listed. (B)

(22) Sustainable forms of land use should be developed within the primary economic sector. In particular, land use should take the biosphere reserve's zonation into account. (B)

(23) In the secondary economic sector (crafts, industry) energy consumption, use of raw materials and waste management should be oriented to guidelines for sustainable development. (B)

(24) The tertiary economic sector (services inter alia in retail, transport and tourism) should be oriented to guidelines of sustainable development. (B)

■ Ecosystem energetics and landscape management

(25) The aims, concepts and measures for protection, maintenance and development of ecosystems and ecosystem complexes, and for regeneration of impaired areas, must be described and implemented. (B)

(26) Animal and plant communities, and their habitat conditions, must be documented, taking into account species and biotopes listed in Red Data Books. Measures for conservation of species that are typical of relevant ecological regions, and habitat-development measures, must be described and implemented. (B)

(27) When interventions are made in ecosystem energetics and in a landscape's appearance, and when compensation and replacement measures are carried out, applicable regional guidelines, and environmental quality targets and standards, must be properly taken into account. (B)

■ Biodiversity

(28) Important sites for floral and faunal genetic resources must be named and described; suitable measures must be designed and implemented for conserving theses resources at the places where they are found. (B)

■ Research

(29) Applied, implementation-oriented research must be carried out within the biosphere reserve. The biosphere reserve itself must provide the database for research on the basis of the AG CIR (1995) ecosystem-type key. The emphases of, and financing for, the research must be documented within the application for designation and in the framework concept. (B)

(30) Third-party research of relevance to the biosphere reserve should be co-ordinated, harmonised and documented by the biosphere reserve's administration. (B)

■ Integrated Monitoring

(31) Proof must be furnished that the necessary staffing, equipment and financial basis is in place to carry out Integrated Monitoring in the biosphere reserve. (B)

(32) Integrated Monitoring in the biosphere reserve must take into account the overall approach to environmental monitoring in Germany's biosphere reserves, as well as the environmental monitoring programmes and concepts of the EU, of the Federal Government and of the Länder, and the existing routine monitoring programmes of the Federal Government and of the *Länder*. (B)

(33) Data relative to the establishment and operation of national and international monitoring systems, and whose generations is required by the MAB programme, must be made available by the biosphere reserve's administration, free of charge, to institutions named for this purpose by the Federal Government and the *Länder*. (B)

■ **Environmental education**

(34) Environmental education topics must be developed within the framework concept, taking into account the biosphere reserve's specific structures, and then implemented in the biosphere reserve. The application must include documentation of environmental education measures, which are a central administrative task. (B)

(35) Each biosphere reserve must have at least one information centre with a full-time staff present throughout the year. The information centre should be supplemented by non-central information offices. (B)

(36) Close co-operation should be sought with existing institutions and educational organisations. (B)

■ **Public relations and communications**

(37) The biosphere reserve must engage in efficient public relations, carried out on the basis of a defined concept. (B)

(38) As part of public relations for a biosphere reserve, consumers and, especially, product manufacturers, must be encouraged to support economically viable, sustainable development. (B)

(39) The services of advisers ("moderators") should be used to promote communications among users and to facilitate the balancing of interests. (B)

(...)

German MAB National Comittee (Ed.) (1996): Criteria for Designation and Evaluation of UNESCO Biosphere Reserves in Germany, Bonn.

Hint:
Free copies of the complete text can be ordered at:
MAB-Geschäftsstelle im BfN
Konstantinstr. 110
D-53179 Bonn

Further information and up-to-date order sheets at:
www.unesco.org/mab and www.biosphärenreservate.de

7.2 *List of Abbreviations*

AGBR/EABR: Exchange of experience in the biosphere reserves in Germany (Erfahrungsaustausch der Biosphärenreservate Deutschlands); formerly: Working Group of the Biosphere Reserves (Arbeitsgruppe der Biosphärenreservate)
ATKIS: Official Topographical-Cartographic Information System (Amtliches Topographisch-Kartographisches Informationssystem)
BfN: Federal Agency for Nature Conservation (Bundesamt für Naturschutz)
BMU: Federal Ministry for the Environment, Nature Conservation and Nuclear Safety (Bundesministerium für Umwelt, Naturschutz und Reaktorsicherheit)
BNatSchG: Federal Nature Conservation Act (Bundesnaturschutzgesetz)
BR: Biosphere Reserve
BRIM: Biosphere Reserves Integrated Monitoring
BSE: Bovine Spongiform Encephalopathy
BUND: German Association for the Environment and Nature Conservation (Bund für Umwelt und Naturschutz Deutschland e.V.)
BVVG: Land-privatisation Agency of the Federal Government (Bodenverwertungs- und -verwaltungsgesellschaft mbH)
CBD: Convention on Biological Diversity
CCC: Climate Change Conference
CCD: Convention to Combat Desertification
CI: Conservation International
CITES: Convention on International Trade in Endangered Species of Wild Fauna and Flora
COP: Conference of the Parties
DBU: German Federal Foundation for the Environment (Deutsche Bundesstiftung Umwelt)
DDT: Dichlorodiphenyltrichloroethane
DLG: German Agricultural Society (Deutsche Landwirtschaftsgesellschaft)
DVL: German Association for Landscape Management (Deutscher Verband für Landschaftspflege e.V.)
EMAS: Eco-Management and Audit Scheme
T&D: Trial and development project
FCCC: Framework Convention on Climate Change
FFH: EU Habitats Directive
FSC: Forest Stewardship Council
GEF: Global Environmental Facility
GIS: Geographical Information System
GTZ: German Technical Cooperation (Gesellschaft für Technische Zusammenarbeit GmbH)
LU: Livestock unit
IBA: Important Bird Area
ICC: International Coordination Council
ICLEI: International Council for Local Environmental Initiatives

IUCN: The World Conservation Union; formerly: International Union for Conservation of Nature and Natural Resources
LABO: Federal Government/*Länder* Working Party on Soil Conservation (Bund/Länder-Arbeitsgemeinschaft für Bodenschutz)
LANA: Federal Government/*Länder* Working Party on Nature Conservation, Landscape Management and Recreation (Bund/Länder-Arbeitsgemeinschaft für Naturschutz, Landschaftspflege und Erholung)
LAWA: Federal Government/*Länder* Working Party on Water (Bund/Länder-Arbeitsgemeinschaft Wasser)
LN: Agricultural land (Landwirtschaftliche Nutzfläche)
LNatSchG: Land Nature Conservation Act (Landesnaturschutzgesetz)
LSG: Landscape Conservation Area (Landschaftsschutzgebiet)
MAB: Man and the Biosphere
MODAM: Multi-Objective Decision Support Tool for Agroecosystem Management
NABU: German Association for Nature Conservation (Naturschutzbund Deutschland e.V)
NLP: National Park
NN: Sea level (Normal Null)
NQZ: Nature Conservation Quality Objectives (Naturschutz-Qualitätsziele)
NRP: Nature Park
NGO: Non-Governmental Organisation
NSG: Nature Conservation Area (Naturschutzgebiet)
OECD: Organization for Economic Cooperation and Development
ÖLV: Biological Food and Marketing Co. (Ökologische Lebensmittel GmbH & Co Vermarktungs KG)
QM: Quality management
PPF: Peace Park Foundation
R&D: Research and development project
spp.: species pluralis (Latin), several species
spec.: species (Latin)
SRU: The German Advisory Council on the Environment (Rat der Sachverständigen für Umweltfragen)
TK: Topographical map (Topografische Karte)
TLG: Governmental Real Estate Trust (Treuhandliegenschaftsgesellschaft GmbH)
TMAP: Trilateral Monitoring and Assessment Program
UFOPLAN: Federal Environment Ministry's environmental research plan
UNCED: United Nations Conference on Environment and Development "Rio Conference 1992"
UNESCO: United Nations Educational, Scientific and Cultural Organization
UVP: Environmental impact assessment (Umweltverträglichkeitsprüfung)
WBGU: Federal Government Scientific Advisory Committee on Global Environmental Change (Wissenschaftlicher Beirat der Bundesregierung Globale Umweltveränderungen)
WHO: World Health Organization
WWF: World Wide Fund for Nature (formerly: World Wildlife Fund)
WTO: World Trade Organisation
WTO: World Tourism Organisation

7.3 Glossary

Abiotic: without life, lifeless
Acidification: change in the pH value (unit of measurement of the existing hydrogen ions) in the acidic range

Active Regions: nationwide competition awarded by the Federal Ministry for Consumer Protection, Food and Agriculture in 2002

Adaptive management: ecosystemic processes are often non-linear and time-delayed. Adaptive management is capable of addressing such uncertainty factors and nevertheless reaching rational decisions. For example, it entails learning phases and phases in which feedback from research is awaited.

Age Class Management: managing the forest by dividing it into areas that are each completely harvested and then replanted, so that all of the trees in an area are always of the same age

Age Class Management: managing the forest by dividing it into areas that are each completely harvested and then replanted, so that all of the trees in an area are always of the same age

Agriculture, biological-dynamic: land use according to the rules of organic farming and taking account of cosmic influences according to Rudolf Steiner

Agriculture, conventional: conventional land use with the goal of largely maximising yields; use of fertilisers and plant protection agents in conventional cultivation systems ("good technical practice" and "integrated land management") according to the existing legislative situation (Fertilisers Ordinance, Plant Protection Act)

Agriculture, integrated: cultivation system under conventional agriculture with the goal of minimising synthetic aids by using environmentally friendly methods (damage threshold principle: if exceeded, chemical-synthetic biocides are used); no uniform or binding requirements going beyond existing legislation

Agriculture, organic: characterised by dispensing with readily soluble mineral fertilisers and synthetic plant protection agents; there are defined and binding cultivation guidelines for organic farming that are laid down by the organic associations or in Regulation (EEC) 2092/91 (EC Organic Farming Regulation); compliance with the requirements is monitored

Agro-foresting: frequent form of management found in rain forest areas combining agricultural and forestry usage patterns; traditionally, it has an important role in development cooperation projects; e.g. cultivation of fruit trees in forests to reduce pest infestation; moves away from destructive slash and burn

Anthropogenic: created by man or originating or changing under his influence

Audit: checking quality and environmental management systems within a company; management instrument for the systematic, documented and objective identification of deviations in the actual state of an audited area from the defined targets

Balje: wide and deep water course in the mudflat area that is directly connected to the open sea or connected by means of Seegats

Barrier island: island made of sand, parallel to the coast, with large dunes

Best Practice: best realised solution

Biodiversity: diversity of ecosystems, biotic communities, species and genetic variation within a species

Biological automation: natural processes and self-regulating mechanisms of nature should be exploited as far as possible in forest management

Biosphere: totality of the part of the earth inhabited by living creatures

Biotic: referring to living creatures, to life

Biotope: habitat characterised by certain plants and animal communities

Biting off: eating buds, shoots and leaves of young trees by unguents

Bodden: shallow, irregularly formed bay with a narrow opening to the sea

Boreal: living in northern regions

Bottom-up process: process in a hierarchical system, going from the lowest structural level to the top structural level via various intermediate levels (opposite: top-down process)

Brushwood fence: low groyne-like dam made of placed bushes or placed rockfill; encourage water calming by means of silt deposits and, thus, land reclamation on the coast

Clearing House Mechanism: instrument to spread information and "know how" to implement the CBD

Conservation: one of the functions of UNESCO Biosphere Reserves (cf International Guidelines, Article 3)

Contractual nature conservation: contracts with land users under conditions of management compatible with nature or nature conservation, often with specific objectives for species and biotope conservation; financial compensation for a reduction in income

Crop rotation: sequence of various crops in a field; growing changing crops according to certain principles; prevents soil fatigue and the spread of pests, but is also necessary for farm organisational and economic principles

Cultural landscape: landscape developed over history due to use by man and shaped by the usage forms with largely anthropogenic ecosystems (unlike a natural landscape)

Degressive funding: funding with the amount of funding falling over time

Deutsches MAB-Nationalkomitee: German MAB National Committee

Ecosystem complex: composition of several ecosystem components, therefore to be evaluated more highly at a spatial level and at the level of interlinking

Ecosystem: according to Article 2 of the CBD, a dynamic complex of communities of plant, animal and microorganism communities and their non-living environment interacting as a functional unit

Ecosystems, accumulating: habitats in which organic substances in particular accumulate due to decelerated material conversion processes; under humid climatic conditions these in particular include natural moor locations with their peat storage (carbon accumulation)

Ecosystems, semi-aquatic: habitats with permanent water saturation of the site (up to a shallow flood) and swamp vegetation adapted to this; usually sites with accumulation of organic substance

Endemites: species of plants and animals that are native only to a very limited area

Eulittoral: periodically dry area on the coast; tidal zone between the level of the mean high tide and the mean low tide

Eurytopic: not tied to certain environmental conditions; common

Eutrophication: accumulation of nutrients that lead to changes in an ecosystem or parts thereof

Fauna: totality of wild animal species

Field: arable field; in crop rotation the arable land is divided into single fields: corn field, maize field, etc.

Fischer-Tropsch method: method for producing fuels from coal

Flora: totality of wild plant species

Fraying: in game with antlers (stag, roebuck), rubbing the fully formed antlers on trunks and branches with the velvet being removed from the antlers

Functions of a UNESCO Biosphere Reserve: conservation, use and logistical support (cf International Guidelines, Article 3)

Groyne: dam built out into the water from the shore

Habitat: location where an animal or plant species regularly occurs

Hydromelioration: extreme drainage of a site used for agriculture or forestry with the objective of raising the yield potential and/or ease of working; usually implemented with large-scale exclusion of ecological aspects

Hypertrophication: extreme over-supply of nutrients

IUCN Protected Area categories: division of worldwide protected areas into (value-free) categories depending on main goal of protection: Strict Nature Reserve/Wilderness Area (I a), Wilderness Area (I b), National Park (II), Natural Monument (III), Habitat/Species Protection Area (IV), Protected Landscape/Protected Sea Area (V) and Resource Protection Area (VI).

Land maintenance: collective term for the subjects of nature conservation and landscape management including open space planning

Land/Länder (pl.): Federal State. The Federal Republic of Germany consist of 16 Federal *Länder.*

Landscape Framework Plan: landscape planning at the regional level (e.g. governing district, region, district), beside others as a specialist contribution to the regional plan (cf Article 15 BNatSchG)

Landscape management: preparation and implementation of measures to safeguard sustainable usability of natural commodities as well as the diversity, uniqueness and beauty of nature and the landscape

Large-scale protected areas: collective term for biosphere reserves, nature parks and national parks

LEADER II: EU Community initiative within the framework of the Structural Fund for Rural Development; encouraging the economic development of rural communities in the least developed regions

LIFE: since 1992, EC finance instrument for pilot projects in the fields of environment, nature and third countries

Logistical support: one of the functions of UNESCO Biosphere Reserves (cf International Guidelines, Article 3)

Major nature conservation projects: projects in the Federal funding programme for "Establishing and Safeguarding Parts of Nature and the Landscape worthy of Protection that have National Representative Significance" (since 1979) and in the Federal Water Edge Project (since 1989)

Melioration: measures for the permanent improvement of the usability of the soil for agriculture and forestry

Monitoring: long-term, regularly repeated and targeted surveys along the lines of permanent observation with statements on the state of and changes to nature and the landscape

Natura 2000: pan-European system of protected areas as an implementation of the Habitat and Bird Protection directives

Natural: unchanged by man, in original condition

Nature Rangers: collective term for full-time and voluntary protected area workers with a monitoring and information function

Neozooa/Neophytes: migrated or introduced (or accidentally introduced) animal and plant species alien to the area that form self-sustaining populations in the newly occupied area
Paludiculture: used ecosystems in which highly productive biomass can be produced under semi-aquatic site conditions

Peeling: chewing off as yet not very thick tree bark by unguents as part of their diet

Placed bushes: groups of shrubs or brushwood; is permanently installed between a double row of posts in brushwood fences and secured with wire to stop it floating away

Placed bushes: groups of shrubs or brushwood; is permanently installed between a double row of posts in brushwood fences and secured with wire to stop it floating away

Placing a value on: economic use of a previously unused resource to make a profit ("turning into money"); term not used uniformly, sometimes also synonymous with "valuation"
Planning procedures: legally binding procedure in which all interests are to be weighed up in order to grant permission to build

Polder: area in the flood plain of a water course surrounded by dykes that is supposed to act as flood protection

Polytrophication: over-supply of nutrients; here it applies to a site, an ecosystem usually characterised by species-poor, highly competitive and highly productive vegetation

Pontic species: species originating from the Black Sea area

Predators: animal species that feed on other animals (unlike herbivores)

Process conservation: allowing all of the natural, both biotic and abiotic, processes for the ecosystem concerned

Public-Private Partnership: equal cooperation of private and public establishments (e.g. limited company – local authority)
Ramsar Convention: agreement on wetlands, in particular as a habitat for wading and water birds, of international importance; designated Ramsar areas

Red Lists: lists of endangered species, species communities and biotopes at Land, national, European or international level
Resources: stocks of a material and ideal nature that are usually available only to a limited extent

Rough grazing: unfenced, extensive grazing area on which farm animals are grazed

Rural: of the countryside, rustic

Seegat: channel between neighbouring barrier islands; combines the mudflat area with the open sea

Segetal flora: from the Latin "segetalis", belonging to the seed; flora that accompany arable crops; includes plants that form communities with crops; their cultivation is beneficial or vital for them (also arable weeds)

Semi-natural: developed without direct human influence and not greatly changed by man, close to the natural state

Stakeholder: player, involved party

Succession: chronological sequence of species or biotic communities in the development of a biotope

Thermophile: loving the warmth

Top-down process: process in a hierarchical system, going from the top structural level to the bottom structural level via various intermediate levels (opposite: bottom-up process)

Trophy: nutrient supply/content of an ecosystem

Ubiquitous: occurring everywhere, represented everywhere

Unguents: wild cloven-hoofed animals subject to hunting law, including the deer family (cervidae) with the native representatives red deer and roe deer as well as the non-native representatives fallow deer and sika deer, the bovine family (bovidae) with the representatives chamois, mouflon and goats, as well as pigs (wild boar)

Use: one of the functions of UNESCO Biosphere Reserves (cf International Guidelines, Article 3)

Valuation methods: methods for estimating or calculating the value of a previously unused resource. For example, a familiar method is an analysis of willingness to pay: surveys among target groups reveal how much they would be willing to pay to use a resource – or to stop using a resource.

Valuation: allocation, estimation or calculation of a value for a previously unused resource. In most cases this value is expressed monetary terms, but is can also be of a non-monetary nature. A monetary value estimated or calculated in a valuation does not have to be implemented in practice.

Xylobiont: living off/in wood

7.4 *Subject Index*

7.5 List of Authors

Baranek, Elke:
Engineer in landscape planning and garden design; freelance moderator with the emphases on participation processes and participatory procedures in land use planning, policy-making, urban and regional development as well as consumer research; freelance activity, among others for Bornholdt-Ingenieure GmbH, Scientific Assistant at the Technology and Society Centre at Berlin Technical University

Blahy, Beate:
Engineer for Veterinary Medicine; employee of the Schorfheide-Chorin BR administration; key subjects: public relations, nature rangers and international relations; Chair of the Federal Association of Nature Rangers (Bundesverband Naturwacht e. V.)

Boley-Fleet, Kirsten:
Geographer; Head of the Nature Conservation Department at the Land Office for the National Park and the Schleswig-Holstein Wadden Sea BR administration; key subjects: regional protection concepts, MAB, regulations on intervention, cooperation with associations and local authorities

Brendel, Ulrich:
Biologist; employee of Zukunft Biosphäre GmbH - Gesellschaft zur nachhaltigen Entwicklung mbH in the Berchtesgaden BR; key subjects: concepts for sustainable development, species conservation, ecological assessments, environmental education

Brendle, Uwe:
Administrator; Head of the Department "Nature and Society", Federal Agency for Nature Conservation (BfN); Member of Advisory Council "Active Regions"

Dittrich, Monika:
Geographer; employee of the German Technical Cooperation (Deutsche Gesellschaft für Technische Zusammenarbeit (GTZ) GmbH); key subject: management of natural resources

Druckrey, Frauke Dr:
Chemist; 1992-2001 environmental expert at the Verband der Chemischen Industrie e. V. (VCI); key subjects: Responsible Care, sustainable development and basic issues of environmental policy; Member of the German MAB National Committee, Member of the International UNESCO MAB Focus Group on Quality Economies in Biosphere Reserves

Engels, Barbara:
Biologist; Scientific Officer at the Divisions "International Nature Conservation/MAB" and "Recreation, Leisure Use and Tourism", Federal Agency for Nature Conservation (BfN)

Erdmann, Karl-Heinz Dr:
Geographer; Head of the Division "Social and Legal Foundations of Nature Conservation", Federal Agency for Nature Conservation (BfN); on the teaching staff at the University of Bonn and Vechta University

Frielinghaus, Helmut:
Agricultural engineer; Head of Division "Cultivation" at Ökodorf Brodowin Landwirtschafts GmbH & Co. KG

Gätje, Christiane Dr:
Head of the Department "Monitoring and Research" at the Land Office for the National Park and the Schleswig-Holstein Wadden Sea BR administration; key subjects: coordination of ecosystem research projects and Wadden Sea monitoring, setting up and implementing socio-economic monitoring, sustainable tourism

Geier, Michael:
Landscape maintenance engineer, landscape gardener; Director of the Bavarian Administrative Office of the Rhön BR

Gietl, Susanne:
Forestry engineer and teacher; Pedagogical Director of the "Wildniscamp am Falkenstein", Bavarian Forest BR/NLP

Göppel, Josef:
Graduate forestry engineer; Member of the German Parliament (Bundestag) (Chair of the Environment Working Group of the CSU), Chair of the Deutscher Verband für Landschaftspflege (DVL), Member of the Supervisiory Board of the Deutscher Naturschutzring, Member of the German MAB National Committee

Grimm, Johannes Dr:
Agricultural scientist and farmer; Projekt Management Dr Grimm (agriculture, ecology, environment, science), currently Managing Director of Naturschutzhof Brodowin

Günther, Beate:
Biologist and German language specialist; freelance activity and scientific assistance as mediator (BM), moderator, trainer, supervisor in design and participation processes in the field of public planning; project manager in numerous environment projects as well as Agenda 21 processes in Berlin and Brandenburg

Hain, Benno Dr:
Biologist; Head of the Division "Basic Questions of Ecology", Federal Environmental Agency (UBA); key subjects: environmental quality objectives, Convention on the Alps, Convention on Biological Diversity and Ecosystem Research; Head of the working group "Mountain-Specific Environmental Quality Objectives" of the Convention on the Alps

Hatzfeldt, Hermann Graf:
Economist; proprietor of the Hatzfeldt-Wildenburg Administration, Schloss Schönstein (Wissen); Chair of the FSC Working Group Germany, Member of the Council for Sustainable

Development, Member of the German MAB National Committee

Hein, Gertrud Dr:
Geographer; Head of Department for further training at the Nature Conservation and Environmental Protection Academy NRW (NUA) set up at the Land Institute for Ecology, Soil Management and Forestry of the Land NRW (LÖBF); Member of the German MAB National Committee

Heinze, Birgit:
Licentiate in Natural Sciences (TR), translator, mediator; Associate Scientific Officer at the division "International Nature Conservation/MAB", Federal Agency for Nature Conservation (BfN); key subjects: finance instruments, EU enlargement, Middle East, MAB; MAB Secretariat at the Federal Agency for Nature Conservation (BfN)

Hellmuth, Elke Dr:
Graduate teacher and geographer; Deputy Director of the Vessertal-Thuringian Forest BR; key subjects: landscape planning and land use

Henne, Eberhard Dr:
Veterinary surgeon; Director of the Schorfheide-Chorin BR, 1st Chair of EUROPARC Deutschland

Heyne, Peter:
Biologist; Director of the Upper Lausitz Heath and Pond Landscape BR

Höltermann, Anke Dr:
Graduate forester; Scientific Officer at the division "Integrative Nature Conservation in Agriculture, Forestry and Settlements", Federal Agency for Nature Conservation (BfN)

Jarmatz, Klaus:
Biologist; Director of the Schaalsee BR

Job-Hoben, Beate:
Ecologist; Scientific Officer at the division "Recreation, Leisure Use, Tourism", Federal Agency for Nature Conservation (BfN)

Kehl, Christine Dr:
Biologist; project manager in the water margin project in the Spree Forest BR

Königstedt, Brigitte Dr:
Biologist; Scientific Officerat the Elbe River Landscape BR, Lower Saxon Section; key subjects: species conservation concepts and biotope management

Kruse-Graumann, Lenelis Prof Dr:
Psychologist; Head of the Teaching Area "Ecological Psychology" at the Institute for Psychology, Hagen Correspondence University; Honorary Professor at Heidelberg University; key subjects: environmental psychology as well as social and linguistic psychology, member of numerous bodies, including the Federal Government Scientific Advisory Council on Global Environment Change (to 2000) and the German MAB National Committee

Kullmann, Armin:
Agricultural engineer; Head of the Section "Regional Advice and Marketing", Institute for Rural Structural Research at Frankfurt a. M. University (IfLS)

Mack, Rolf-Peter Dr:
Agro-biologist; Director of the Project "People and Biodiversity in Rural Areas. Sustainable Livelihoods for Diverse Cultural and Natural Landscapes" of the German Technical Cooperation (Deutsche Gesellschaft für Technische Zusammenarbeit (GTZ) GmbH)

Mattern, Kati:
Geographer; Scientific Officer at the division "Basic Questions of Ecology", Federal Environmental Agency (UBA); key subjects: ecosystemic environmental monitoring, indicators, environmental quality objectives and Convention on Biological Diversity

Mayerl, Dieter:
Landscape maintenance engineer, landscape gardener; Director at the Bavarian Environment Ministry (1974-2003); key subjects: landscape planning/maintenance, including for biosphere reserves; Member of the German MAB National Committee (1991-1995); Spokesman of the Working Group for UNESCO Biosphere Reserves in Germany (AGBR) (1997-2000)

Meier, Ariane:
Communications engineer; post-graduate studies in the faculty for Environment, Culture and Society at the University of Lancaster, Institute for Environment, Philosophy and Public Policy (GB); key subject: cultural perception of nature

Nauber, Jürgen:
Graduate forester; Head of the Division "International Nature Conservation/MAB", Federal Agency for Nature Conservation (BfN); Secretary-General of the German MAB National Committee

HAN Nianyong:
Biologist; Secretary-General of the Chinese MAB National Committee; Chinese Academy of the Sciences (Beijing)

d'Oleire-Oltmanns, Werner Dr:
Biologist; Managing Director of Zukunft Biosphäre GmbH - Gesellschaft zur nachhaltigen Entwicklung mbH in the Berchtesgaden BR

Petschick, Michael:
Agricultural engineer; Head of the Section "Ecologisation of Land Use in the Spree Forest BR; key subjects: LEADER project development, regional management, contractual nature conservation

Plachter, Harald Prof Dr Dr h. c.:
Biologist; Professor for Nature Conservation in the Biology Faculty at Marburg University; key subjects: population ecology of endangered species, nature conservation significance of natural disruptions (flooding, grazing), nature conservation in agricultural landscapes, nature conservation working methods (evaluation, planning), general strategies of nature conservation; Member of the World Commission on Protected Areas of the IUCN, the German delegation on the UNESCO World Heritage Convention and the German MAB National Committee as well as numerous expert bodies (including Federal Research Ministry, Deutsche Bundesstiftung Umwelt), among others

Pokorny, Doris Dr:
Land maintenance engineer; Deputy Director of the Bavarian Administrative Office of the Rhön BR; key subjects: coordination of research and environmental monitoring, international cooperation; Member of the UNESCO Advisory Committee for Biosphere Reserves

Precht, Folkert Dr:
Geographer; Head of the Division "Sciences" of the German UNESCO Commission (Deutsche UNESCO Kommission, DUK)

Preyer, Rolf-Dieter:
Administrator; Senior Officer in the Division "Research and Technology Policy of the Federal Foreign Office; key subject: UNESCO scientific programmes

Puhlmann, Guido:
Engineer in melioration; Director of the Middle Elbe River Landscape BR administration

Reiter, Karin Dr:
Biologist; Scientific Officer at the division "Integrative Nature Conservation in Agriculture, Forestry and Settlement", Federal Agency for Nature Conservation (BfN)

Remmers, Irmgard:
Land maintenance engineer; Director of the National Park Administration Lower Saxon Wadden Sea (also competent administration for the Lower Saxon Wadden Sea BR)

Rimpau, Jürgen Prof Dr:
Agricultural scientist; member of numerous bodies, including the Board of the Deutsche Landwirtschafts-Gesellschaft e. V., the Council for Sustainable Development and the German MAB National Committee

Schönthaler, Konstanze:
Land maintenance engineer; partner and holder of special statutory authority at Bosch & Partner GmbH; key subjects: ecosystemic environmental monitoring, environmental quality objectives, indicators, protected area planning as well as Convention on the Alps and Convention on Biological Diversity

Schreiber, Hans-Joachim:
Graduate agricultural engineer; Head of the Department "Nature Conservation and Landscape Maintenance", Environment Ministry of the Land Mecklenburg-Vorpommern; Spokesman for UNESCO biosphere reserves in Germany, Member of the German MAB National Committee

Schulz, Christiane:
Geo-ecologist; thesis and project support "Sustainable Land Use" in the Spree Forest BR administration in the section "Ecologisation of Land Use"

Schulz, Werner F. Prof. Dr:
Economist; Professor for Environmental Management at the University of Hohenheim, Institute for Business Management and Director of the German Skills Centre for Quality Economies (DKNW) at Private University Witten/Herdecke gGmbH; Member of the German MAB National Committee

Specht, Rudolf Dr:
Biologist; Scientific Officer at the division "International Nature Conservation/MAB", Federal Agency for Nature Conservation (BfN)

Steinmetz, Elke:
Geographer and biologist; Scientific Officer at the division "General and Fundamental Issues in Nature Conservation", Federal Ministry for the Environment, Nature Conservation and Nuclear Safety (BMU); key subject: MAB

Succow, Michael Prof Dr:
Biologist; Director of the Botanical Institute and the Botanical Garden, Greifswald University; key subjects: plant ecology in the field of moor ecology, vegetation-oriented landscape ecology and nature conservation; Winner of the Alternative Nobel Prize 1997, Chair of the Foundation Council of the Michael Succow Foundation for Nature Conservation, Member of the German MAB National Committee

Treß, Johannes:
Agricultural engineer; Director of the Vessertal-Thuringian Forest BR

Vogel, Michael Dr:
Biologist; Director of the Berchtesgaden National Park; President of the Alpine Protected Areas Network

Walter, Alfred:
Economist; Head of the Division "General and Fundamental Issues in Nature Conservation", Federal Ministry for the Environment, Nature Conservation and Nuclear Safety (BMU); Chair of the German MAB National Committee

Weigelt, Michael Dr.:
Biologist; Director of the Rügen National Park Office (responsible for the South East Rügen BR and Jasmund National Park)

Wenzel, Peter:
Graduate forester; Head of the Department "Nature Conservation and Forests", Ministry for Agriculture and Environment of the Land Saxony-Anhalt